Thonner's analytical key to the families of flowering plants

Leiden Botanical Series, Volume 5

Series ISBN 90 – 6021 – 462 – 5

Leiden Botanical Series is a publication of Leiden University Press, The Hague/Boston/ London

In the Leiden Botanical Series will be published papers of a monographic nature from the entire field of botany (including its history, bibliography, and biography) which by their length (100 printed pages or more) are unsuitable for publication in journals. Information can be obtained from the editors. Rijksherbarium, Schelpenkade 6, Leiden, The Netherlands.

Thonner's analytical key to the families of flowering plants

R. Geesink
A. J. M. Leeuwenberg
C. E. Ridsdale
J. F. Veldkamp

Springer-Science+Business Media, B.V. 1981

R. Geesink – Rijksherbarium, Leiden, Netherlands

A. J. M. Leeuwenberg – Laboratorium voor Plantensystematiek en Planten-
geografie, Agricultural University, Wageningen, Netherlands

C. E. Ridsdale – B. A. Krukoff Botanist of Malesian Botany, Rijksherbarium,
Leiden, Netherlands

J. F. Veldkamp – Rijksherbarium, Leiden, Netherlands

This volume is listed in the Library of Congress Cataloging in Publication Data

This is a translated and revised edition of: Anleitung zum Bestimmen der Familien der
Blütenpflanzen, 2nd. ed. 1917, Friedländer, Berlin

ISBN 978-94-010-9892-2   ISBN 978-94-010-9890-8 (eBook)
DOI 10.1007/978-94-010-9890-8

Contents

'All plants are hybrids, but some
are greater bastards than others'

Franz Thonner. 1910. Photo L. Grillich, Vienna. Original in Library of the National Botanical Garden of Belgium.

Preface to the 2nd edition (1917)

While most European floristic works contain keys for the identification of families, genera, and species, extra-European ones usually have the keys replaced by a systematic survey, which makes it cumbersome to identify the family to which the plant under investigation belongs. It is, of course, often possible to recognize it immediately by the presence of a conspicuous character, but there are also many cases in which it is not that easy, and then written aid such as given in the present work is desired. The few works of this kind presently available generally consider the typical features of the family only, while they neglect the numerous exceptions. In the present work, however, especially in the second edition, all exceptions have been considered as far as possible, the less significant ones in footnotes,[1] so that plants with characters that are different from the ones typical for the family may also be identified correctly.

In the choice of differentiating characters those have been preferred that can be seen in a flowering plant with the naked eye. As far as nomenclature, delimitation, and description of the families are concerned Engler and Prantl's *Die natürlichen Pflanzenfamilien* has been used as the basis for the revision of the present work; in addition, however, many other manuals have been consulted, especially Bentham and Hooker's *Genera plantarum*, De Candolle's *Prodromus*, Baillon's *Histoire des plantes*, and Engler's *Pflanzenreich*, as well as various floras.

The key is followed by a description of all families of flowering plants,[2] which mainly serves to check the result of the identification for its correctness, as well as an explanation of the most important botanical technical terms occurring in the book. The addition of figures has been decided against; one finds suitable ones especially in Engler's *Syllabus der Pflanzenfamilien*.

Vienna, in May 1917   **Franz Thonner**

---

1   Incorporated into the key in the current edition.
2   Omitted in the current edition.

Introduction

For the identification of a flowering plant the first step usually is to discover to which family it belongs. With some experience, the families commonly encountered in one's area of interest are soon known, but when dealing with specimens from other places, notably those from the vast and rich subtropics and tropics, there is much less certainty. The pertinent literature is often not readily available as it is often found only in expensive, rare or obscure books, or journals, present only in a few specialized institutes.

Basically only a few keys to the families of flowering plants of the world have ever been produced, the best known of which at present is Hutchinson's *Key to the families of flowering plants* (1973); less well-known are Lemée's *Tableau analytique des genres monocotylédones* (1941) (incl. *Gymnosperms*) and his *Tableau analytique des genres dicotylédones* (1943), and Hansen and Rahn's *Determination of Angiosperm families by means of a punched-card system (Dansk Bot. Ark. 26,* 1969, with additions and corrections in *Bot. Tidsskr. 67,* 1972, 152–153, and *Ibid. 74* 1979, 177–178). Of note also are Davies and Cullen's *The identification of flowering plant families,* 2nd ed. (1979), which, however, deals only with the families native or cultivated in North Temperate regions, and Joly's *Chaves de identifição das famílias de plantas vasculares que ocorrem no Brasil,* 3rd ed. (1977), which may be useful in other tropical areas too.

There are a number of excellent keys prepared by an Austrian, Franz Thonner (1863–1928), which deal either with European genera (1901, 1903, 1918), or African ones (1908, 1913, 1915), or with all families of the world (1891, 1895, 1917). Some of these have apparently been completely overlooked, others have been known only to a few, and then sometimes served as a base for keys of their own, thereby again influencing keys by others (see *Derived works*).

At Dutch Universities extensive use has long been made of the *Anleitung zum Bestimmen der Familien der Blütenpflanzen,* 2nd ed. (1917, Friedländer, Berlin), which to our experience has proven to be the most reliable work in existence. Of course, as the keys deal with a highly complex subject, they require close attention for a profitable use. They may therefore perhaps have scared off even professional botanists, who then had to take recourse to other simpler and therefore less dependable ones. In a few places, Thonner's keys were better appreciated and even introduced in undergraduate courses, for instance by Pulle and his school in the Netherlands, by Sørensen in Copenhagen, and in Brazil at first by an unknown translator and later by Rawitscher, Alvim and Joly. Elsewhere the *Anleitung (1917)* has been little known, rare and, for many, inaccessible, as it is in German.

It seemed, therefore, a worthwhile venture to translate it into English. A start was made by Leeuwenberg in the early 1960s, but other obligations soon delayed progress. About twelve years later, he mentioned this in a casual conversation with Geesink and Ridsdale, who had just begun a translation of their own, and his efforts were thankfully incorporated. Veldkamp joined shortly afterwards. It rapidly became apparent that mere translation would be unsatisfactory: the innumerable footnotes should also be worked into the main key; the nomenclature should be brought up to date; and something should be done about the many new families accepted by some authors since Thonner's time. For the latter we have largely restricted ourselves to those mentioned by Airy Shaw in his revision of Willis' *A dictionary to the families of the flowering plants and ferns,* 8th ed. (1973) and Hutchinson's *The families of flowering plants,* 3rd ed. (1973), these being currently the most consulted manuals. These 'segregated' families have now all been accounted for.

We have also tried to check the many curious or aberrant genera, but have undoubtedly missed many. The keys have not become easier because of all these additions. The number of key couplets has increased from 812 (excluding footnotes) in the 1917 edition, to 2117 in the present one. Nevertheless, they provide a useful means of identification and force students as we know from experience, to make a clear and careful analysis and logical interpretation of the various parts of the plant. We hope that all those interested through profession or hobby may be aided in a rapid identification of their material, and that we have made Franz Thonner and his works slightly better known and appreciated.

We invite the user to point out errors, difficulties, and omissions. It should then be indicated in which couplets difficulties arose with a suggestion as to how they might be remedied. A representative specimen would be useful, even if only on loan. Any assistance will be acknowledged in future editions. Communications should be sent to R. Geesink or J. F. Veldkamp, Rijksherbarium, Schelpenkade 6, P.O. Box 9514, 2300 RA Leiden, the Netherlands.

Leiden, September 1980

Acknowledgements

Thonner spent about 30 years creating his *Anleitung (1917)*, apparently without much outside help. We were more fortunate and had others to advise and assist us. First of all we thank the Director, Staff, and students of the Rijksherbarium, Leiden, for providing the facilities, expert knowledge, and trial runs of the key, respectively. Other help was promised by many, but given by few. We had many helpful suggestions and criticisms but have applied the remarks in our own fashion, hence all mistakes and misinterpretations made should be attributed to us. Our sincere thanks are due to R. C. Bakhuizen van den Brink Jr. (Leiden, *various*), M. M. J. van Balgooy (Leiden, *Elaeocarpaceae, various*), G. M. Barroso (Rio de Janeiro, *Lepidocordia*), B. G. Briggs (Sydney, *Proteaceae, Restionaceae*), R. Clarysse (Meise, Thonner/De Wildeman correspondence), M. J. E. Coode (Kew, *Elaeocarpaceae*), T. A. Cope (Kew, *various*), T. B. Croat (Saint Louis, *Araceae*), P. J. Cribb (Kew, *Orchidaceae*), R. Dahlgren (Copenhagen, esp. *Monocotyledones*), F. G. Davis (Kew, *Compositae*), J. Dransfield (Kew, *Palmae*), L. L. Forman (Kew, *Fagaceae*), P. S. Green (Kew, *Oleaceae*), C. Grey-Wilson (Kew, *Balsaminaceae*), B. Hansen (Copenhagen, *Balanophoraceae, various*), C. Hansen (Copenhagen, *Melastomataceae*), R. M. Harley (Kew, *Labiatae*), P. Hiepko (Berlin, *Opiliaceae*), Ding Hou (Leiden, *Anacardiaceae, Aristolochiaceae, Celastraceae, Hippocrateaceae*), S. S. Hooper (Kew, *Cyperaceae*), D. R. Hunt (Kew, *Commelinaceae*), B. R. Jackes (Atherton, *Epacridaceae, Vitaceae*), L. A. S. Johnson (Sydney, *Gymnospermae*), Hsuan Keng (Singapore, *Gymnospermae*), R. Kool (Leiden, *Ixonanthaceae*), K. U. Kramer (Zürich, *various*), J. Kuyt (Lethbridge, *dicotyledonous parasites*), D. J. de Laubenfels (Syracuse, *Gymnosperms*), P. W. Leenhouts (Leiden, *Burseraceae, Connaraceae, Sapindaceae*), D. J. Mabberley (Oxford, *Adoxaceae, Meliaceae, Sterculiaceae*), W. Marais (Kew, *Chloanthaceae, Liliaceae*), W. Margadant (Utrecht, *biohistory of Thonner*), S. Mayo (Kew, *Araceae*), J. F. Maxwell (Singapore, *Melastomataceae*), N. L. Menezes (São Paulo, *Joly key*), R. van der Meijden (Leiden, *Haloragaceae, Polygalaceae*), H. P. Nooteboom (Leiden, *Simaroubaceae, Symplocaceae*), W. R. Philipson (Christchurch, *Calycanthaceae, Idiospermaceae, Monimiaceae*), P. H. Raven (Saint Louis, *promotion in the U.S.A.*), J. W. A. Ridder-Numan (Leiden, *various small families*), R. E. Rintz (Mt. Clemens, *Asclepidiaceae*), M. J. Sands (Kew, *Balanitaceae, Begoniaceae*), M.

Schmid (Noumea, *New Caledonian taxa*), C. G. G. J. van Steenis (Leiden, *Bignoniaceae, Sonneratiaceae, various*), B. C. Stone (Kuala Lumpur, *Pandanaceae, Rutaceae*), M. Tamura (Osaka, *Ranunculaceae*), N. P. Taylor (Kew, *Cactaceae*), B. N. Teensma (Leiden, *Portugese*), J. Thompson (Sydney, *Tremandraceae*), C. C. Townsend (Kew, *Amaranthaceae*), P. van der Veken (Gent, *various*), W. Vink (Leiden, *Hamamelidaceae, Sapotaceae, Winteraceae*), E. F. de Vogel (Leiden, *Apostasiaceae, Orchidaceae, seedlings*), J. N. Westerhoven (Hirosaki, *Ikeno key*), W. J. J. O. de Wilde (Leiden, *Myristicaceae, Najadaceae, Passifloraceae*), K. L. Wilson (Sydney, *Cyperaceae, Juncaceae*).

We thank the Botanical Garden, Berlin, for the opportunity to show a poster there during its tercentenary celebration in September 1979. We assume that at least those who ran off there with a free copy of the Preliminary Version (or obtained one later) but never bothered to comment have found it to be without blemish.

We are most obliged to P. W. Leenhouts, Leiden, who was willing to assist us in correcting the proofs and who painstakingly checked the numbering again.

The reproductions of the pictures of Thonner were made by B. N. Kieft and the drawings for the plates by J. van Os, Leiden.

Finally, we thank our wives, who first had to miss us on Thursday evenings ('*Thonnerstagabend*'), and later had to spend holidays during which manuscripts were polished and retyped, but never complained too much.

Franz Thonner – Life (1863 – 1928)

Franz Thonner was born in Vienna on 11 March 1863 as the son of Franz Thonner, cordwainer at the Imperial Court of Vienna, and Therese Schnaubelt. Very little is known of his life. Most of the following has been extracted from the sources mentioned below, which usually give only the briefest information.

He was educated at the Theresien Gymnasium in Vienna, and then studied Law for a single semester (in Vienna ?). His interest then turned to the Natural Sciences, to which he remained devoted for the rest of his life. He studied in Vienna and Berlin, but apparently never obtained an academic degree. In 1891 he married Marie Svoboda, a Czech; there is no record of any children. They first settled in Dresden, but in 1903 moved to Vienna, where they remained until 1920. Afterwards they went to Smichov, a suburb of Prague, where Thonner died on 21 April 1928.

Somehow Thonner was a gentleman of private means, which allowed him to pursue the subjects of his interest and thus became what in German is called a 'Privatgelehrter'. It is remarkable that he turned to larger projects only, at least only one brief article (1897) from his hand is known to us. When only 28, he had already written and published a key to the families of flowering plants of the world, the *Anleitung (1891)*, a unique work, as no one before him had prepared a similar treatise. He paid for this publication himself, as he did for all his subsequent ones. The absence of an experienced publishing house perhaps explains why his works remained almost unnoticed in the scientific journals of that time and they remained virtually unnoticed to the present day. Possibly to increase his market and also to include his later additions, he translated them into French or English, in which languages he was well versed. For further details see the next chapters on *Bibliography* and *Derived works.*

Together with his wife he often travelled through Europe and North Africa. Twice he went on his own to the Ubangi and Mongala Districts of the Belgian Congo. Both expeditions were cut short: the first (23 August – 22 October 1896) because the Congolese went off with his canoe and some of his equipment and collections; the second (28 January – 16 March 1909) because of illness, so he collected much less than he had intended.

He wrote journals on each expedition in German (1898, 1910) and in French (1899, 1910), which contain a wealth of orginal botanical, ethnological, and linguistic observations. About the botanical collections, two books were also written, for the publication of both of which he also paid (De Wildeman & Durand, 1900; De Wildeman, 1911). In the first book, De Wildeman observed that although only 120 botanical collections were made 50 were new for the area, and 23 species and 4 varieties were new to science. It is rare that such a proportion would be obtained; he apparently had a keen eye and had gone well prepared. In the second book, De Wildeman took the opportunity to publish extensively on the flora and vegetation of the area, an action heartily approved of by Thonner.

Several of the new species were named after him, but unfortunately the only genus named in his honour, *Thonnera* De Wild. (*Annonaceae*), has turned out to be a synonym of Uvariopsis Engl. & Diels (see *Eponymy*).

Next to nothing is known about his private life and methods. He apparently rarely visited the Naturhistorisches Hofmuseum in Vienna (Thonner, *in litt.*, K.-H. Rechinger, Vienna, *pers. comm.*) mainly to check identifications and to select material for his illustrations. He probably corresponded with the Botanisches Museum in Berlin, since he asked De Wildeman to send duplicates of his collections to Diels, Engler, and Harms, but the Berlin archives were destroyed during World War II. We procured part of his correspondence, mainly with De Wildman in Brussels (March 1899 – May 1921), from which some information could be gleaned. Although the two must have known each other for a long time, met occasionally and visited the Opera together, the brief notes remain formal. Their wives corresponded also; how tantalizing to know more of what they had to tell each other! Thonner's handwriting was even and clear, as is shown by the accompanying sample (p. xvi and xvii), one of the few where mention is made of the *Anleitung (1917)*.

For his plates he privately employed an artist, J. Fleischmann, who was for a short time assisted by another one, not named, who made the analytical drawings. At least one of his manuscripts, written by him in stenography, was worked up to a definitive version by an unknown secretary.

To us his major works are the various keys to the genera and families. Although we have studied the *Anleitung (1917)* for a long time now, we can still only guess about his methods. Each of his keys was basically different from the preceding ones, as may be noted from the main couplets, a change which necessitates an entirely new structure. He apparently based himself especially on Engler and Prantl's *Die natürlichen Pflanzenfamilien (1895–1915)* and *Das Pflanzenreich* as far as it had appeared, as can be seen from the sometimes verbatim quotations. It is interesting to note that many genera originally misplaced there key out in the *Anleitung (1917)* to the families where they have subsequently been transferred to. Whether he had an extensive file or a prodigious memory we do not know, but the results speak for themselves: they have never been surpassed.

In 1911, he was awarded a Belgian distinction, apparently at the request of De Wildeman, but as yet we have not discovered which nor the citation of the award.

During World War I he sent part of his private library to Great Britain as a payment for the publication of *The flowering plants of Africa (1915)*, as transfer of funds was prohibited. After the War, his fortunes dwindled with the incredible inflation of those times, and he wrote that he tried to subsist by translating novels between English, French and German. His correspondence, if any, with De Wildeman after 1921 is lacking from the archives of Brussels.

Of his last years in Smichov, we know nothing, except that he fell victim to a chronic disease and died on 21 April, 1928 at the age of 65.

The only obituary that we have received (through the kind efforts of the librarian of the Naturhistorisches Museum, Vienna) was in a Viennese anthropological journal; to the botanical world he remained virtually unknown both in life and death.

Sources

Anonymous. (1928). Enciclopedia Universal Ilustrada Europeo-Americana 61: 678. Madrid.

Degener, H. A. L. (1922). Wer ist's? (Unsere Zeitgenossen): 1569. Leipzig.

De Wildeman, E. A. J. (1899–1921). Personal correspondence with F. Thonner. Msc. Brussel.

Schmid, B. & C. Thesing. (1914). Biologen-kalender 1: 325. Berlin.

Želízko, J. V. (1928). Franz Thonner. *Mitt. Anthrop. Ges. Wien*: 238. Wien.

Franz Thonner – Bibliography

1891. Anleitung zum Bestimmen der Familien der Phanerogamen. viii+280 p. Friedländer, Berlin.

1895. Analytical key to the natural orders of flowering plants. vii+151 p. Swan Sonnenschein, London, Macmillan, New York.

1897. Das Gebiet des Mongalaflusses in Centralafrika (Kongostaat). Globus 72: 117–121, 7 f.

1898. Im Afrikanischem Urwald. Meine Reise nach dem Kongo und der Mongalla im Jahre 1896. x+117 p., 86 t., 3 m. Reimer, Berlin.

1899. Dans la grande forêt de l'Afrique Centrale. Mon Voyage au Congo et à la Mongala en 1896. x+115 p., 87 t., 30 f., 3 m. Schepens & Cie, Bruxelles.

1900. Introduction, in: De Wildeman, E. & Th. Durand, Plantae thonnerianae congolenses. xi–xx, 23 t., 1 m. Schepens & Cie, Bruxelles.

1901. Exkursionsflora von Europa. x+50+356 p. Friedländer, Berlin. (Reprint, 1980, Rijksherbarium, Leiden).

1903. Flore analytique de l'Europe. vi+322 p. Baillière, Paris.

1908. Die Blütenpflanzen Afrikas. Eine Anleitung zum Bestimmen der Afrikanischen Siphonogamen. xvi+672 p., 150 t., 1 m. Friedländer, Berlin.

1910. Vom Kongo zum Ubangi; meine zweite Reise in Mittel-Afrika. xi+116 p., 114 f., 3 m. Reimer, Berlin.

1910. Du Congo à l'Ubangi; mon deuxième voyage dans l'Afrique Centrale. xii+126 p., ill., 1 m. Misch & Thron, Bruxelles, Rivière, Paris. (n.v.).

1911. Introduction, in: De Wildeman, E., Plantae thonnerianae congolenses. II. ix–xvii, 20 t., 51 f., 1 m. Misch & Thron, Bruxelles.

1913. Die Blütenpflanzen Afrikas. Nachträge und Verbesserungen. 88 p. Friedländer, Berlin.

1915. The flowering plants of Africa. An analytical key to the genera of African phanerogams, xvi+647 p., 150 t., 1 m. Dulau, London. (Reprint, Hist. Nat. Clas. 27, 1962, Cramer, Weinheim).

1917. Anleitung zum Bestimmen der Familien der Blütenpflanzen, ed. 2. vi+280 p. Friedländer, Berlin.

1918. Exkursionsflora von Europa. Nachträge und Verbesserungen. 55 p. Friedländer, Berlin. (Reprint, 1980, Rijksherbarium, Leiden).

Wien IV. Paniglgasse 20, am 1.I.1917

Sehr geehrter Herr Dr Wildeman!

Nach langer Zeit erlaube ich mir wieder einmal
anzufragen, wie es Ihnen geht und Ihnen gleichzeitig
unsere besten Glückwünsche zum Jahreswechsel
zu übermitteln.

Bei uns geht alles so ziemlich seinen
gewohnten Gang. Eigentlich spüren wir nicht
viel vom Krieg und leben fast wie vor demselben.
Den Sommer haben wir teils in Baden bei Wien,
teils in Plana in Böhmen zugebracht. Die englische
Ausgabe meines Werkes über die afrikanischen
Pflanzen ist nun endlich erschienen, durch den

Krieg verzögert, aber, dank der Ermittlung eines schweizer Bekannten, nicht verhindert. Eine neue Auflage meines ersten Werkes (Bestimmungs, tabellen für Pflanzenfamilien), die mich in den letzten Jahren beschäftigt hat, wird demnächst in Druck gehen.

Indem ich Sie bitte, Ihrer werten Frau Gemahlin und Fräulein Tochter meine und meiner Frau herzlichste Glückwünsche zum Neuen Jahre übermitteln zu wollen, verbleibe ich

Ihr ergebener

Franz Thonner.

Franz Thonner – Derived works

Thonner's efforts remained more or less unknown. Two botanical works were based on his expeditions to the Belgian Congo (*De Wildeman & Durand, 1900; De Wildeman, 1911*), of which the first sold only 4 copies in the first year (he gave away a number as complimentary copies). Apparently his two journals did not fare much better, but were perhaps of sufficient importance as an obituary appeared in an anthropological journal. On his keys a few others were directly or indirectly based, and are listed here. Possibly there are more, of which we would like to be notified; they can easily be detected by the sequence of the main couplets, if no mention is made in the introduction.

In 1893 Ikeno published an abbreviated Japanese translation of the *Anleitung (1891)*.

Henriquez (1897) translated it into Portuguese, but the journal in which it appeared did not have a wide circulation, and this translation was for instance apparently unknown in Brazil.

Pittier translated the Analytical key (1895) into Spanish and adapted it for use in South America. The first edition (1917) was used by Standley (1920), who was apparently unaware of its Anglo-American origin, for his Mexican keys. Standley used the second edition of Pittier's *Clave* (1926) for his Panaman flora (1928). A third edition appeared in 1939.

Joly (1977) discussed in length the discovery in 1939 of a manuscript key in use in Viçosa, Brazil, which turned out to be derived also from the *Analytical key (1895)*. This key was mimeographed several times before it was revised by Rawitscher and Rachid-Edwards (1956), and again independently revised and restricted to Brazil by Alvim (1943) and Joly (1969).

We ourselves also distributed a stencilled Provisional Edition (1979) of 106 copies to various institutes and colleagues for comment.

Alvim, P. De Tarso. (1943). Chave para a determinação das famílias das plantas Pteridophytas, Gimnospermas e Angiospermas brasileiras ou exóticas encontradas no Brasil. (Adaptação da chave de Franz Thonner . . .). Msc. Viçosa. (*Fide Joly, 1977*). Reprint (1950) 61 p. Viçosa.

Anonymous. (Undated). Title unknown ('Chave . . .'). Msc. 91 p. Escola de Minas da Universidade do Brasil. Several mimeographed editions. Viçosa, later also in São Paulo. (*Fide Joly, 1977*).

Geesink, R., A. J. M. Leeuwenberg, C. E. Ridsdale & J. F. Veldkamp. (1979).

Thonner's analytical key to the families of flowering plants. Provisional version. Stencil. Unpaged (173). Rijksherbarium, Leiden.

Henriquez, J. A. (1897). Clave para a determinação das famílias das plantas Phanerogamicas. *Bol. Soc. Brot.* 14: 82–160.

Ikeno, S. (1893). Kenka-shokubutsu-bunka Kensaku-hen.

Joly, A. B. (1968–1977). Botânica chaves de identifição das famílias de plantas vasculares que ocorrem no Brasil, baseadas em chaves de Franz Thonner. Provisional version (1968); 1st ed. (1970) iv+132 p.; 2nd ed. (1975) 159 p.; 3rd ed. (1977) 159 p. Companhia Editora Nacional, São Paulo.

Pittier, H. (1917–1939). Clave analítica de las familias de plantas fanerógamas de Venezuela y partes adyacentes de la América tropical. Ed. 1 (1917) vii+108 p.; Clave analítica de las plantas superiores de la América tropical. 2nd ed. (1926) viii+130 p.; 3rd ed. (1939) vii+94 p. Lit. y Tip. del Comercio, Caracas.

Rawitscher, F. K. & M. Rachid-Edwards. (1956). Title unknown. (*Fide Joly, 1977*).

Standley, P. C. (1920) Key to the families, in: Trees and shrubs of Mexico. *Contr. U.S. Nat. Herb.* 23: 19–36. Washington.

—. (1928). Key to the families, in: Flora of the Panama Canal Zone. *Contr. U.S. Nat. Herb.* 27: 50–65. Washington.

Wildeman, E. A. J. de. (1911). Etudes sur la flore des districts des Bangala et de l'Ubangi (Congo Belge). Plantae thonnerianae congolenses. II xvii+465 p., 20 t., 1 f., 1 m. Misch & Thron, Bruxelles.

Wildeman, E. A. J. de. & Th. Durand. (1900). Plantae thonnerianae congolenses, ou énumération des plantes récoltées en 1896 par Fr. Thonner dans le district des Bangalas. ix+49 p., 23 t., 1 m. Schepens & Cie, Bruxelles.

Franz Thonner – Eponymy

A number of taxa collected by Thonner have been named after him. Leeuwenberg was able to consult the original set in BRUX and based on this the following list could be compiled. An asterisk indicates that the name has been considered as correct in recent revisions. Unmarked ones for which no synonymy is given have not recently been treated as far as known.

Thonnera De Wild. (Annonaceae) = Uvariopsis Engl. & Diels

Aframomum thonneri De Wild. (Zingiberaceae)
Antholyza thonneri De Wild. (Iridaceae) = Gladiolus atropurpureus Bak.
Bertiera thonneri De Wild. & Th. Dur. (Rubiaceae)*
Casearia thonneri De Wild. (Flacourtiaceae) = C. barteri Mast.
Clerodendrum thonneri Gürke (Verbenaceae)*
Combretum thonneri De Wild. (Combretaceae) = C. paniculatum Vent.
Conopharyngia thonneri (Stapf) Stapf (Apocynaceae) = Tabernaemontana thonneri De Wild. & Th. Dur. ex Stapf
Crotonogyne thonneri De Wild. (Euphorbiaceae) = C. poggei Pax
Dichapetalum thonneri De Wild. (Dichapetalaceae) = D. bangii (F. Didr.) Engl.
Dicranolepis thonneri De Wild. & Th. Dur. (Thymelaeaceae) = D. buchholzii Engl. & Gilg
Dinophora thonneri Cogn. (Melastomataceae) = Phaeoneuron dicellandroides Gilg
Dioscorea thonneri De Wild. & Th. Dur. (Dioscoreaceae) = D. preussii Pax
Harveya thonneri De Wild. & Th. Dur. (Scrophulariaceae)*
Hygrophila thonneri De Wild. (Acanthaceae)
Impatiens thonneri De Wild. & Th. Dur. (Balsaminaceae) = I. irvingii Hook. f. ex Oliv.
Isolona thonneri (De Wild. & Th. Dur.) Engl. & Diels (Annonaceae)*
Listrostachys thonneriana Kränzl. (Orchidaceae) = Diaphananthe pellucida (Lindl.) Schltr.
Loranthus thonneri Engl. (Loranthaceae) = Agelanthus brunneus (Engl.) v. Tiegh.
Macaranga thonneri De Wild. (Euphorbiaceae) = Alchornea laxiflora (Benth.) Pax & Hoffm.
Millettia thonneri De Wild. (Leguminosae)*

Monodora thonneri De Wild. & Th. Dur. (Annonaceae) = Isolona thonneri
Engl. & Diels

Ouratea thonneri De Wild. (Ochnaceae)*

Pycnocoma thonneri Pax (Euphorbiaceae)*

Rhabdophyllum thonneri (De Wild.) Farron (Ochnaceae) = Ouratea thonneri
De Wild.

Rinorea thonneri De Wild. (Violaceae) = R. welwitschii (Oliv.) O. Ktze

Rourea thonneri De Wild. (Connaraceae) = Roureopsis thonneri Schellenb.

Roureopsis thonneri (De Wild.) Schellenb. (Connaraceae)*

Scaphopetalum thonneri De Wild. & Th. Dur. (Sterculiaceae)*

Sesamum thonneri De Wild. & Th. Dur. (Pedaliaceae) = ? S. mombazense De
Wild. & Th. Dur.

Tabernaemontana thonneri De Wild. & Th. Dur. ex Stapf (Apocynaceae)*

Thunbergia thonneri De Wild. & Th. Dur. (Acanthaceae)

Uragoga thonneri De Wild. & Th. Dur. (Rubiaceae) = ? Psychotria sp.

Urera thonneri De Wild. & Th. Dur. (Urticaceae)*

Vitex thonneri De Wild. (Verbenaceae)

The Key – Introduction and Notes

Each of Thonner's keys was different from the preceding ones. As the present work was initially intended to be a mere translation of the *Anleitung (1917)*, we have not changed its structure, even when some major couplets are notoriously difficult. On the whole Thonner has managed to keep the key as simple as possible, and so have we; but highly technical questions which need some botanical experience and a good dissecting microscope cannot be avoided. Some will therefore find it a difficult book to use at first. We would suggest some methods to facilitate use.

Start with some well-known plants, or back-track your way from a few familiar families; in this way, you will become acquainted with the keys and the terms used. It will then be noted that they are based on relatively few characters which turn up time and again. Unfortunately complete material is required: sterile and exclusively male specimens cannot be identified, female or fruiting ones will cause great problems. For these, Hansen & Rahn's punch-cards will limit your options.

The key is strictly dichotomous (except for some couplets in the *Concise key to the groupings*): each couplet is composed of two leads. The latter are usually composed of two parts again, separated by a dash. The first part should be contradicted by the opposing lead of the couplet. The second part contains additional information; features mentioned here may or may not be present in taxa referable to the opposing lead; they are merely given as a possible further aid. In both parts the characters are given in the morphological descriptive sequence, if feasible, and not according to their diagnostic 'weight'. This has been done to facilitate reading; many keys have been made more difficult and confusing because of their scrambled text. Distribution is often also given as an aid, but is of course only valid for plants not introduced, cultivated, or escaped; especially weedy or showy plants should be suspect, while exact distributions are still not always known in some cases.

Read both leads carefully and completely!

Try to visualize their intentions and use your brains! Most misidentifications are due to careless, hasty, sloppy, superficial, and unimaginative reading. Note the numbers encountered on a slip of paper, marking uncertain choices to facilitate retracing if you go wrong.

Do not pick and poke about the specimen!

The various leads are in a haphazard morphological sequence and you should try to limit destruction of your specimen as much as possible; once it has been torn apart it will be difficult to reconstruct and you may need another flower of your precious material!

Boil a single flower!

You can always boil another if required. Fresh material is often easier to handle after boiling, too. Examine it in a Petri-dish under sufficient water so that it will neither float away, nor be obstructed by the surface of the water; a drop or two of detergent will drive off air bubbles (chaffy flowers as in *Cyperaceae*, *Gramineae* do not need to be boiled at all, some detergent in water is sufficient); soak overnight in strong ammonia when the floral parts are flimsy and glued together, as in *Balsaminaceae* and *Orchidaceae*.

Make a short diagnosis!

It is often useful to do so, working from the outside inwards in such a way that nothing is inadvertently damaged that may be needed later, for instance after you have found the correct family and have to use the material with other keys all over again. See the accompanying scheme as a guide (p. xxvi). Simple sketches will also be helpful, for instance a floral diagram (aestivation!) and shape of fragile parts.

Add these notes and sketches, and as much as can be saved and dried of the remnants of the object to the specimen for future reference.

A difficult question was how to mention the many new families accepted by some since Thonner's time. We have largely limited ourselves to those in Willis' *Dictionary* (1973) and Hutchinson's *Families* (1973). In some cases, we are convinced that their distinction is unacceptable, in others that they are indeed distinct, but in many cases, as in the *Liliales*, *Saxifragaceae s. l.* few specialists agree. So who are we to profess expertise to make a satisfactory choice among the options? As this key is primarily intended for practical use, and not as a taxonomic manual, we thought we should have some leeway; in principle we decided to follow Melchior's *Engler's Syllabus der Pflanzenfamilien, 12th ed. (1964)*, but deviated from this course where it suited us. It was also borne in mind that Thonner himself based his family concept on Engler and Prantl's *Pflanzenfamilien*. One should therefore not invoke our arbitrary use of names in an argument on the taxonomic distinctness of such a family. The fact that supposedly related taxa often key out close together should not be extrapolated to doubtful cases, as the keys are artificial; such coincidences are merely fortuitous (yet, there may be something in it, one never knows!).

The segregated families are noted in brackets as in the *Exkursionsflora (1901)* and other works. Genera and some supra-generic taxa have been noted when we had the impression that these would key out exclusively in a particular lead, but only when one or two taxa seemed to be involved, e.g. '(*Escalloniaceae:Itea*) . . . **Saxifragaceae**'. This means that only *Itea* keys out here, which is sometimes treated as an *Escalloniacea*, which family is treated here as part of the *Saxifragaceae*. Some notes of warning: a taxon may well turn up in several places without being noted everywhere, partly because of the artificiality of the key (an apetalous species will end up in a different place from its petaliferous congeners), whereby it may run down together with more than two other taxa in places, partly because we overlooked it. More taxa than those mentioned may actually key out to one place, but we were not aware of it. The taxon may not belong here at all (we hope not), but was included because of an error by us, or because the descriptions in the literature consulted were faulty (by necessity we had to lean heavily upon other works). We are convinced that not all aberrant taxa have been included, partly because we simply were not aware of their existence, partly because the conventional, less controversial, and often huge families such as the *Euphorbiaceae, Myrtaceae*, and those of the *Tubiflorae* have been much less studied.

Some taxa may appear to have been misplaced in the key but are not the result of a misinterpretation. Instead, their 'wrong' inclusions act as fail-safes, many of which were already built into the system by Thonner in his footnotes. In several instances, features are not what they seem to be, but this is then only known to someone familar with the situation, who will then not use these keys in the first place. Bracteoles may be adnate to a perianth and then resemble a calyx, suggesting a place among the *Chori-* or *Sympetalae*; petals may be so cohesive that they appear connate and mislead the unsuspecting to the *Sympetalae*, on the contrary they may be fused at the very base only, appearing free, suggesting a place among the *Choripetalae*. As this key aims to be practical, we have maintained despite objections from some learned correspondents, that the plant should also key out according to the interpretation of the structure which would appear most logical to someone not hampered by knowledge, even if this is morphologically incorrect.

Thonner's keys were rarely illustrated and more plates in the current work would have been useful, but as we wanted to remain as concise as possible, we have refrained from adding more. One is therefore referred to the other works mentioned by Thonner and in our introduction, and to the many other text-books. For world families. Heywood's recent *Flowering plants of the world (1978)* provides an inexpensive and well-illustrated survey.

The terms employed will usually cause no great difficulty. We have tried to use as few technical terms as possible, including those required in the *Glossary* at the end of the book, sometimes ad absurdum; for those we missed one should consult Jackson's *A glossary of botanic terms, 4th ed. (1928)*. We hope to have solved the problem about *hypo-*, *peri-*, and *epigyny* by the footnote to Couplet 548 and by Plate 1, while the most common types of ovules have also been depicted (Plates 2 and 3). One ambiguous term has been pointed out by various colleagues which we refuse to change: *epipetalous* (or *-tepalous*) means *'opposite to the petals (or tepals), but* **not** *necessarily inserted on them'*. Others use these words to indicate insertion only, and not relative position, whereby the term *alternipetalous* (or *-tepalous*) has no uninomial, easy counterpart.

Thonner included short descriptions of the families and they are indeed very useful for speedy reference. We had to omit these at present and the user is referred to other manuals. It was not possible to prepare reliable succinct diagnoses, even when so many are available. To copy these from existing literature proved unsatisfactory, as descriptions are often not complete enough to fit the *Scheme for a diagnostic description* as is given on page xxvi, a most surprising discovery. Their deletion has one minor advantage to the buyer of this book: it would otherwise have been much thicker and more expensive.

Scheme for a diagnostic description

Note position, number, coherence, shape, and size where applicable.

Vegetative characters
Habitat (*if not terrestrial*).
Life form (*annual, perennial, shrub, tree, climber, liana*).
Indument (*check young parts*), type of hairs.
Leaves (*arrangement, simple/compound, type of nervation*), presence of trans-
  lucent lines or dots, crystals (*strong pen light useful here, mind your eyes!*),
Stipules (*absence/presence, check young shoots, scars*).

Floral characters
Inflorescence (*type, mode of branching*); bracts; bracteoles.
Flower (*sex, actino-/zygomorphic, hypo-/peri-/epigynous, see Plate 1*); aestiva-
  tion (*in bud*) of sepals, petals, tepals; hypanthium.
Disk (*absence/presence; extra-/intra-staminal*).
Stamens (*alterni-/epipetalous or -tepalous*); filaments (*free/ad-/connate*);
  anthers (*dehiscence by slits, pores, valves; in-/la-/extrorse – check in bud*).
Styles; stigmas (*number of lobes may be indicative of number of carpels and
locules*).
Ovary (*superior/(hemi-)inferior – Plate 1*); locules; placentas; ovules (*position,
type, see Plates 2 and 3, number per locule/ovary*).

Fruiting characters
Fruit (*type, dehiscence, consistency*).
Seeds (*number per locule or fruit; surface; appendages and their position*).
Embryo (*form, position: the radicle points to where the micropyle was!*).
Endosperm (*absence/presence, consistency*).

Origin (*only for truly indigenous plants*).

CONCISE KEY TO THE MAJOR GROUPINGS

(N.B. When in doubt consult the main key!)

1

KEY TO THE FAMILIES

1. Reproductive organs ('*flowers*') unisexual, often subtended by bract-like structures, rarely by 2 or 4 free or connate, opposite bracteoles (*Gnetales*), but true perianth absent. Stamens ('*micro-sporophylls*') more or less developed, several to many together ('*pollen cones*' or '*micro-sporangia*'), each with 2–many, rarely 1, anthers ('*pollen-sacs*'). Carpels ('*macro-sporophylls*') not connate into a closed ovary. Ovules naked, rarely enclosed in a utricle, atropous or ana-tropous, sessile, 1–several together, subtended by a bract; bracts usually aggregated into cones. Seeds exposed, or enclosed, either by the bracts ('*cone-scales*') or by parts of the seed-bearing structure ('*epimatia*'), these usually woody or leathery, sometimes fleshy and pseudo-carp berry- or drupe-like, rarely seed more or less enclosed in a basally attached, fleshy aril.—Stem woody. (*Gymnospermae*). 2
— Flowers unisexual or bisexual. True perianth usually present. Anthers usually on a filament. Ovules completely enclosed by the ovary.[1] Fruit very rarely cone-like. Seeds completely enclosed by the fruit, which may dehisce at maturity.—Style usually present. (*Angiospermae*). 16

GYMNOSPERMAE

2. Flowers usually in branched and very compound inflorescences; at least the male ones with a pseudo-perianth of 2 opposite, free and/or 2 more or less connate bracteoles. Resin absent.—Leaves oppo-site, simple, sometimes scale-like. Ovule enclosed in a utricle, atro-pous with a style-like, elongated integument. Seeds nut-like, en-closed by fleshy bracts in a cone or a drupe- or berry-like syncarp. (*Gnetales*). 3
— Flowers solitary, or in capitules, or in spikes, or in cones. Pseudo-perianth absent, flowers usually subtended by bract-like scales. 5

---

1 Incompletely so in *Degeneriaceae, Nelumbo, Platanus, Resedaceae*.

3. Shrubs, trees, or woody climbers with well-developed trunks. Leaves more than 2, scale-like or well-developed, pinninerved..... 4
— Woody perennial with a very stout, truncate, subterraneous stem, apically bi-lobed, each lobe with a strap-shaped, parallel-nerved leaf, which may tear to the base.—Male flowers with 2 free and 2 connate bracteoles, 6 micro-sporophylls at base connate into a tube and a pistillode. Deserts of S.W. Africa........... **Welwitschiaceae**
4. Virgate shrubs. Leaves small, scale-like, connate. Flowers in cones. Male flowers with 2 connate bracteoles and 2–8 micro-sporophylls on an androphore. Warm temperate Eurasia, N. and S. America.
Ephedraceae
— Usually climbing shrubs, rarely trees. Leaves well-developed, free, pinninerved. Flowers whorled in spikes. Male flowers with 2 bracteoles connate into a tube. Micro-sporophylls 1 or 2 on an androphore. Tropics. .... **Gnetaceae**
5. Bole tuberous or columnar, simple, rarely branched, then usually only at the apex and branches not originating from axillary buds. Sap slimy. Leaves large, palm-like, usually accompanied by scales, terminally tufted. Pollen sacs many (25 or more) per micro-sporophyll, in clusters of 2–6. (*Cycadales*)..................... 6
— Bole usually branched, branches slender, originating from axillary buds. Sap usually resinous. Leaves moderately sized to small, not palm-like, simple, entire or fan-shaped. Pollen sacs up to 15 per micro-sporophyll, sometimes in 2 rows, never in clusters......... 8
6. Leaflets either with a midrib and lateral nerves, or without a midrib and nerves parallel. Ovules 2 per bract, floral axis not growing through the female pseudo-flower. 7
— Leaflets with a midrib, lateral nerves absent. Ovules 4–8, rarely 2 per bract. Floral axis growing through the female flower (i.e. the whorls of ovule-bearing bracts). .... **Cycadaceae**
7. Leaflets parallel-nerved, nerves straight or wavy, simple or forked at base. .... **Zamiaceae**
— Leaflets pinninerved, midrib distinct, lateral nerves parallel, forked.—Leaflets convolute in bud. S. Africa........ **Stangeriaceae**
8. Leaves usually with a single midrib, sometimes with additional parallel veins, or a dichotomous venation, rarely with 2 unbranched veins, then leaves in whorls of 16–30, each with a small bract at base (*Sciadopitys*); apex rounded to acute. Ovules either subtended by bracts, or surrounded by sterile parts of a modified shoot structure, almost always by both in a compound structure. 9
— Leaves at base with 2 nerves, which branch dichotomously, midrib absent, apex usually 2-lobed. Ovules usually 2, on a long stalk, each with a cupule at base.—Long and short shoots present. Leaves alter-

nate, long-petioled, broad, fan-shaped. Female inflorescences in the axils of leaf-like bracts. (*Ginkgoales*). .... **Ginkgoaceae**
9. Seed either with a fleshy outer surface, or partly to completely enclosed by a fleshy aril, then drupe-like.—Leaves with a single vein. Ovules atropous, at least partly exposed. (*Taxales*). 10
— Seed rarely fleshy, then ovule anatropous. Fleshy aril absent, but other fleshy structures sometimes present.—Leaves with a single vein, or with a midrib and additional parallel veins. Ovules atropous or anatropous. (*Coniferales*). 11
10. Ovule 1, terminal on a specialized shoot, subtended by several decussate bracts. Seed at least partly enclosed by a fleshy aril, when completely so drupe-like.—Pollen cones and ovule-bearing structures sometimes 2–more together on specialized fertile shoots. **Taxaceae**
— Ovules 2 per bract, axillary; bracts in cone-like inflorescences. Seed with a fleshy outer surface. Pollen structures compound and reduced in cones in the axils of leaves of the preceding year. **Cephalotaxaceae**
11. Ovule 1 per bract. Seed not winged, each surrounded by a fleshy bract, then drupe-like, or bracts forming a fleshy syncarp, or both. Pollen sacs 2 per micro-sporophyll, inverted. 12
— Ovules 1–several per bract. Seed usually winged. Syncarp usually woody, rarely fleshy (*Juniperus*). Pollen sacs 2–more per micro-sporophyll. 13
12. Leaves well-developed or scale-like, entire, phylloclades absent. Ovules usually anatropous, either with a thin cup-like epimatium at base, or enclosed by a leathery or fleshy one, then drupe-like, rarely atropous, then epimatium absent (*Microstrobos*). Pseudo-carp drupe-like. .... **Podocarpaceae**
— Leaves inconspicuous, scale-like, phylloclades present, flabellate, lobed, or dentate. Ovule atropous with a thin epimatium or aril at base. Pseudo-carp a fleshy cone. (*Phyllocladaceae*)... **Podocarpaceae**
13. Pollen sacs usually 3–more per micro-sporophyll, rarely 2. Ovules 1–more per bract, atropous or anatropous. Seed usually with 1–3 wings. Bract adaxially inappendiculate, or with a transverse ridge, or with 1, rarely 2 scales. 14
— Pollen sacs 2 per micro-sporophyll. Ovules 2 per bract, anatropous. Bracts paired, the two more or less free from each other, the outer usually small and thin, the inner enlarging and finally woody.— Leaves solitary or paired or tufted on specialized short shoots with which they are decumbent. Female bracts in a spiral. ..... **Pinaceae**
14. Leaves usually with 1 midrib, rarely with 2 unbranched main nerves, then in whorls of 16–30 (*Sciadopitys*). Female bracts usually not deciduous, if so, then seeds 2–more per bract and bracts without wing-like margins. Ovules usually more than 1 per bract, atropous

5

or anatropous. Seeds usually with 1–3 wings. 15
— Leaves usually with both a midrib and several to many parallel veins. Female bracts usually deciduous with adnate, not winged seeds and with winged margins, if not deciduous, seed with 1 or 2 wings.—Female bracts in a spiral. . . . . **Araucariaceae**
15. Leaves and female bracts decussate or 3 or 4 in a whorl, never distichous.—Ovules atropous, 1–several per bract. Seed not winged or with 1–3 wings. . . . . **Cupressaceae**
— Leaves and female bracts usually in a spiral, distichous or not; leaves rarely opposite on decussate, specialized branchlets with which they are decumbent (*Metasequoia*), or in whorls of 16–30 (*Sciadopitys*).—Ovules atropous or anatropous. Seed with 1–3 wings. . . . . **Taxodiaceae**
16. (1). Stem in transverse section with scattered vascular bundles. Leaves usually parallel-nerved, rarely reticulately so, or absent,[1] usually narrow, undivided, entire, sometimes with adaxial appendages. Flowers usually 3-merous. Pollen usually monocolpate. Cotyledon usually 1, rarely absent. (*Monocotyledones*). 17
— Stem in transverse section usually with the vascular bundles in a ring. Leaves usually reticulately nerved, rarely both narrow and entire, or absent. Flowers usually 4- or 5-merous. Pollen rarely monocolpate. Cotyledons usually 2,[2] rarely only 1,[3] or absent.[4] (*Dicotyledones*). 158

MONOCOTYLEDONES

17. Perianth absent or indistinct, then limited to small scales or hairs, sometimes replaced by tepaloid appendages of the connective, plants then aquatic. 18
— Perianth well-developed in at least the flowers of one sex, then

---

1 Reticulately nerved in many *Araceae, Dioscoreaceae, Musaceae, Orchidaceae, Taccaceae,* some *Liliaceae s.l.*; absent in *Corsiaceae, Geosiridaceae, Lemnaceae, Triuridaceae,* and some *Liliaceae s.l., Burmanniaceae, Cyperaceae, Juncaceae, Orchidaceae, Restionaceae.*
2 Cotyledons 3 or 4; whorled in *Degeneriaceae, Calycanthaceae (Idiospermum);* and *Opiliaceae.*
3 e.g. in some *Portulacaceae (Claytonia), Gesneriaceae (Monophyllaea), Primulaceae (Cyclamen), Cruciferae (Dentaria), Ranunculaceae (Ficaria), Papaveraceae (Corydalis).*
4 In the seedlings of the 'Barringtonia-', 'Garcinia-' and 'Orobanche-' type (cf. De Vogel, Seedlings of Dicotyledons, 1979).

sepaloid, petaloid, or differentiated into a calyx and a corolla. ... 36
18. Flowers, at least the female ones, in simple, rarely compound spadices, which are usually surrounded by a sheath; bracts and bracteoles absent.—Ovary 1.................................... 19
— Flowers not in spadices. 24
19. Terrestrial, rarely fresh-water plants....................... 20
— Submerged, marine plants.—Spadix flattened, consisting of 2 rows
of pairs of 1 stamen and 1 ovary. Anthers extrorse. Pollen filiform.
Ovule 1. Endosperm absent. (*Zosteraceae*)....... **Potamogetonaceae**
20. Flowers bisexual or monoecious, if dioecious leaves dissected.—
Embryo usually large. 21
— Flowers dioecious. Leaves undivided.—Woody plants. Leaves
parallel-nerved, usually tristichous, narrow, margin spiny. Male
inflorescences usually compound. Ovule 1, laterally inserted, or
more. Embryo small............................. **Pandanaceae**
21. Inflorescences simple. Flowers not enclosed by empty bracts. Ovules
1–many, free.. 22
— Flowers enclosed by empty bracts. Ovule 1, completely adnate with
the ovary.—Flowers bisexual or monoecious, then male inflorescences compound. (*Coicineae, Zeeae*)............... **Gramineae**
22. Flowers hypogynous, sometimes immersed in the axis, or with
numerous hairs at base, bisexual or monoecious, then with the male
flowers in the upper part, the female ones in the lower part of the
spadix... 23
— Flowers perigynous, rarely epigynous, monoecious, male and female
flowers alternating in groups or layers in the same spadix.

Cyclanthaceae

23. Leaves distichous, sessile, linear, undivided, entire, parallel-nerved.
Male and female inflorescences separated at least initially by a
bract. Testa dry.—Herbs from marshes or aquatics. Perianth usually
substituted by hairs. Anthers with longitudinal slits. Fruit dry.

Typhaceae

— Leaves in a spiral, usually petiolate, blades sometimes reticulately
nerved, sometimes divided. Male part of the inflorescence when
separate from the female part never subtended by a bract; bracts
and bracteoles absent. Testa fleshy. .... **Araceae**
24. Plant differentiated into stems and leaves. 25
— Plant not differentiated into stem and leaves.—Aquatics, plants consisting of leaf- or grain-like, floating or submerged fronds. Flowers
in depressions of the frond, in groups of 1 pistil (female 'flower')
and 1 or 2 stamens (male 'flowers').................... **Lemnaceae**
25. Ovary 1 and plants submerged marines, or 2–6, collateral, sessile at
least at anthesis and plants aquatics. 26

— Ovary 1, rarely 2–more, then stipitate, usually serial. Terrestrials, or fresh-water aquatics. 30
26. Flowers paired or in spikes, bisexual or polygamous. Stamens 2–numerous. 27
— Flowers solitary or in cymes, monoecious or dioecious. Stamen 1.—Marine aquatics. Style 1, filiform. Stigmas 1–3. (*Cymodoceaceae* sometimes included in *Potamogetonaceae*). . . . . **Zannichelliaceae**
27. Plants of fresh- or brackish-water. Ovaries 3–6. 28
— Marine plants. Ovary 1.—Spikes compound with leaf-like bracts. Stamens 3. (*Posidoniaceae*). **Potamogetonaceae**
28. Flowers several to numerous in simple or compound spikes. Stamens 4–numerous. Fruits subsessile. 29
— Flowers paired. Stamens 2. Fruits finally long-stalked. (*Ruppiaceae*).
Potamogetonaceae
29. Stamens 4, each subtended by a tepal or tepaloid appendage. Ovaries 4. Ovule 1 per ovary, pendulous. Fruits indehiscent.
Potamogetonaceae
— Stamens 6–many, inappendiculate, but 1–3 tepals may be present. Ovaries 3–6. Ovules 2–many per ovary, erect. Fruits dehiscent.
Aponogetonaceae
30. Bracteoles or empty glumes usually present. Filaments well-developed. 31
— Bracteoles absent. Anther 1, subsessile.—Marsh plant. Flowers axillary and in terminal spikes, monoecious, rarely bisexual. Ovary 1. Ovule 1, erect. Style short in the flowers of the spike, very elongated in the basal axillary ones. Endosperm absent. Mountains of Pacific America. (*Lilaeaceae*). . . . . **Juncaginaceae**
31. Flowers solitary, or in simple or compound spikes, or in capitules. Ovules pendulous, 1 per locule or carpel. Fruit a capsule, very rarely indehiscent (?). 32
— Flowers surrounded by membraneous to stiff glumes in variously compound spikelets or pseudo-spikelets, rarely simple, sometimes reduced to 1 flower with some empty glumes. Fruit a caryops, rarely dehiscent. Ovules erect to ascending or completely adnate with the carpel.—Anthers usually 2-locular. 35
32. Terrestrial plants, rarely aquatic, then flowers in capitules. Endosperm present. 33
— Submerged aquatics. Flowers sessile, axillary. Endosperm absent.—Ovary (sub)-sessile. . . . . -. . . . . **Najadaceae**
33. Flowers in capitules, or ovaries several (?). Anthers 2-locular. . . . 34
— Flowers solitary, or in spikelets, or in cymes. Ovary 1. Anthers 1-locular.—Stamens 1 or 2. Ovary 1-locular. Ovule anatropous.
Centrolepidaceae

34. Terrestrials, rarely aquatics, inflorescences then not submerged. Anthers versatile. Ovary one, 2- or 3-locular. Ovule atropous. (*Eriocaulon*)..................................... **Eriocaulaceae**
— Completely submerged aquatics. Anthers adnate. Ovaries (female flowers ?) 1–several, 1-locular. Ovule anatropous.—W. Australia, Tasmania, New Zealand. (*Hydatellaceae*). .... **Centrolepidaceae**
35. Stem usually triangular, solid, without nodes. Leaves at least initially with closed sheaths, ligules often absent. Anthers basifix. Ovule and seed free from the ovary- or fruit-wall, basally attached. Embryo at least partly surrounded by the endosperm.... **Cyperaceae**
— Stem usually terete, hollow, nodose. Leaves with deeply fid sheaths, ligules exceptionally absent, sometimes replaced by a row of hairs. Anthers usually dorsifix. Ovule and seed adnate with the basal lateral side of the ovary- or fruit-wall. Embryo basal, outside the endosperm. (incl. *Anomochloaceae, Bambusaceae, Streptochaetaceae*).
Gramineae
36. (17). Perianth calycoid, sometimes slightly coloured, rarely absent in the flowers of one sex..................................... 37
— Perianth corolloid, or differentiated into a calyx and a corolla.... 82
37. Leaves not both folded in bud and becoming divided later, if so perianth-segments 4 or indistinct and ovules many per carpel..... 38
— Leaves folded in bud, usually becoming pinnately or digitately compound or 2-partite. Perianth-segments usually distinct, then 6 and at least present in flowers of one sex. Ovule 1 per carpel.—Woody plants. Flowers in spatheate spikes, spadices, or panicles. (incl. *Nypaceae*). .... **Palmae**
38. Flowers in spadices with 1–several sheaths.—Fruit indehiscent, or irregularly so, usually fleshy. 39
— Flowers not in spadices. 41
39. Flowers bisexual, monoecious, but then the male ones in the upper part of the spadix and the female ones in the lower. Spadix usually with 1 sheath.—Leaves not plicate. 40
— Flowers monoecious, the male and female ones alternatingly in groups or layers. Spadix with several sheaths.—Leaves 2-partite or flabelliformily partite and/or plicate................. **Cyclanthaceae**
40. Perianth undivided or 4–8-partite. Ovary 1. Fruit a berry, rarely dry and/or irregularily dehiscent........................ **Araceae**
— Tepals 2. Ovaries 3, free. Fruit a follicle......... **Aponogetonaceae**
41. Ovaries inferior or hemi-inferior. 42
— Ovaries completely superior or nearly so, rarely naked. 43
42. Terrestrial plants, or epiphytes. Flowers not spatheate. Perianth-segments 4–6..................................... 77
— Aquatics. Flowers spatheate. Perianth 3-partite.—Flowers solitary

9

or cymosely capitate. Ovary 1-locular. Ovules numerous.

Hydrocharitaceae

43. Ovary 1, 1-locular. 44
— Ovary 1, 2–more-locular, or ovaries 2–more, more or less free. 52
44. Ovule 1.—Herbs with narrow leaves. 45
— Ovules 2–more. 48
45. Flowers solitary, or in pairs, or in fascicles. Endosperm absent. . . 46
— Flowers in spikes, or in capitules, or in panicles. Endosperm present.—Stamens 2–more. Ovule pendulous or descending. 47
46. Male flower with a 2-labiate perianth, the female without any, usually surrounded by a sheath. Stamen 1, anther 1- or 4-locular. Stigmas 2–4. Ovule erect, basal, anatropous. . . . . **Najadaceae**
— Male flower with a cupular perianth or without any, or with one of a few scales, always present in the female flower. Stamens 1–3, sometimes connate, anthers 1- or 2-locular. Ovule apical, pendulous, atropous. . . . . **Zannichelliaceae**
47. Leaves strap-shaped, basal. Flowers in globose capitules, monoecious. Perianth membranous. Stamens 3–more. Ovule anatropous. Fruits more or less drupaceous. **Sparganiaceae**
— Leaves small, scale-like, basal and cauline. Flowers in simple spikes, or in panicles, or in spikelets, usually dioecious. Perianth usually scarious. Stamens 2 or 3. Ovule atropous. Fruit a capsule or a nut.

Restionaceae

48. Leaves petiolate. Perianth-segments 4. Stamens 4. Stigmas 2, sessile. Seeds with a pubescent funicle.—Flowers solitary or in cymes. . . . 49
— Leaves sessile. Perianth-segments 6. Stamens 3 or 6. Stigma 1 or 3 on a simple style. Funicle glabrous. 50
49. Perianth-segments rounded. Ovules apical, more or less anatropous. (*Croomiaceae*). . . . . **Stemonaceae**
— Perianth-segments acute to acuminate. Ovules basal, atropous.

Stemonaceae

50. Stem herbaceous. Leaves not both stiff and serrate. Flowers not in capitules with leaf-like bracts. 51
— Stem woody. Leaves stiff, serrate. Flowers in terminal capitules with leaf-like bracts.—Ovules 2 or 3, basal, erect. Fruit indehiscent. Seed 1. S.W. Australia. (*Dasypogon*). . . . . **Xanthorrhoeaceae**
51. Stigma 1, simple or 3-lobed, not filiform, nor twisted. . . . . . . **Liliaceae**
— Stigmas or styles 3, filiform, twisted. . . . . **Juncaceae**
52. Ovule 1 per locule or free carpel. 53
— Ovules 2–more per locule or free carpel. 55
53. Stamens 1–8(–15). Ovary syncarpous, or free carpels 2–9, rarely numerous, then plants herbaceous, stamens 9, from African marshes. 54

— Stamens and free carpels numerous.—Trees. E. Malesia. (*Sararanga*). . . . . **Pandanaceae**
54. Inflorescences various, if a capitule or a glomerule carpels free. Ovary 2–6-locular, or carpels free, 2–9. Ovules various, if anatropous erect or laterally inserted and ovaries 3–6-locular. 58
— Flowers in capitules without an involucre. Ovary 2-locular. Ovules pendulous, anatropous. . . . . **Sparganiaceae**
55. Ovaries 3–6, free, or connate at base only. 56
— Ovary 1, 3-locular. 72
56. Autotrophic plants of bogs or aquatics. Leaves well-developed. . . . 57
— Saprophytes of tropical forests. Leaves scale-like. (*Petrosaviaceae*).
Liliaceae
57. Herbs of bogs. Flowers in racemes. Tepals 6. . . . . . **Scheuchzeriaceae**
— Aquatics. Flowers in simple or branched spikes. Tepals 1–3.
Aponogetonaceae
58. Ovules pendulous, atropous or hemitropous. 59
— Ovules erect or lateral, anatropous. 68
59. Flowers solitary, paired or in fascicles, axillary. Ovaries free.— Usually marine aquatics with cauline leaves. . . . . . . **Zannichelliaceae**
— Inflorescences otherwise. Ovary 2–4-locular. 60
60. Flowers not in capitules, usually bisexual or dioecious. 61
— Flowers in capitules, usually monoecious.—Perianth present. Stamens 1–4, or 6, free. (*Eriocaulon, Lachnocaulon*). . . . . **Eriocaulaceae**
61. Flowers in umbels, or in spikes, or in panicles. Stamens 4–6(–15). . . . 62
— Flowers in spikelets, arranged into various inflorescences. Stamens 2 or 3. 66
62. Herbs. Leaves parallel-nerved, exceptionally with apical tendrils. Fruit a drupe, or dehiscent into mericarps. 63
— Woody climbers, often with stipular tendrils. Leaves 3–9-plinerved, reticulately viened, petiolate. Fruit a berry. (*Smilacaceae*).
Lilliaceae
63. Flowers in bracteate panicles. Stamens 6. Fruit a drupe. 64
— Flowers in simple spikes. Stamens 4–6. Fruit dry, very spongy, ultimately dehiscent into mericarps. (*Maundia*). . . . . **Juncaginaceae**
64. Erect herbs, without tendrils. 65
— Climbers, often woody at base. Leaves with apical tendrils.—Leaves petiolate, not plicate. Flowers bisexual. Styles 3. (*Flagellaria*).
Flagellariaceae
65. Leaves sessile or very shortly petioled, plicate in bud. Flowers bisexual. Styles (2 or) 3. (*Joinvilleaceae*). . . . . **Flagellariaceae**
— Leaves petiolate, not plicate. Flowers dioecious. Stigma sessile, 3-lobed. (*Hanguanaceae*). . . . . **Flagellariaceae**
66. Anthers 2-locular. Filaments free. 67

— Anthers 1- or 2-locular, then (*Lyginia*) filaments connate at least at base. . . . . **Restionaceae**

67. Radical leaves present, ensiform. Spikelets in spikes or in panicles. Styles 3. Ovary 3-locular. (*Anarthriaceae*). . . . . **Restionaceae**
— Radical leaves absent, cauline ones not ensiform, reduced to scales. Spikelets solitary. Styles 2. Ovary 2-locular. (*Ecdeiocoleaceae*).

Restionaceae

68. (58). Anthers extrorse. Carpels 3 – many, free at least in fruit. Endosperm absent.—Herbs. Leaves ligulate. Flowers sessile, or in spikes, or in racemes, or in panicles. 69
— Anthers introrse or latrorse. Ovary one, 3-locular. Endosperm present. 70

69. Tepals 6. Stamens 4 or 6. Carpels 4 or 6. Embryo straight.—Flowers in spikes or racemes. Stigma sessile. . . . . **Juncaginaceae**
— Tepals either 3 and then stamens 9 and carpels many (*Burnatia*), or 6 and then stamens 3 (*Wiesneria*). Embryo curved. . . . **Alismataceae**

70. Leaves stiff, leathery, serrate or entire. Tepals scarious or bract-like. 71
— Leaves herbaceous, usually entire. Tepals not scarious, nor bract-like. . . . . **Liliaceae**

71. Stem triquetrous, herbaceous. Styles 3, filiform. Exo- and endotesta with a cavity in between. Endosperm mealy. N.E. S. America.

Thurniaceae

— Stem terete, usually woody. Style and stigma 1. Testa without such a cavity. Endosperm cartilaginous. New Guinea to New Zealand.

Xanthorrhoeaceae

72. (55). Style 1. Stigma 1 or 3, rarely styles 3, then not filiform, nor twisted. Endosperm cartilaginous. 73
— Styles or stigmas 3, filiform, usually twisted. Endosperm mealy.— Anthers basifix. 76

73. Plants herbaceous, if woody erect, leaves long-linear, parallel-nerved, flowers in large spiciform panicles and fruit a capsule. . . . 74
— Woody plants, usually climbing and with stipular tendrils. Leaves elliptic to hastate, 3 – 9-pli-nerved, reticulately veined. Flowers small, in umbels, or in racemes, or in panicles. Fruit a berry. (*Smilacaceae*). . . . . **Liliaceae**

74. Stem herbaceous. Leaves not leathery and long-linear, usually entire. Flowers not in large, contracted, spiciform panicles. 75
— Stem usually woody. Leaves stiff, leathery, long-linear, entire to serrate. Flowers small, numerous, in large, contracted, spiciform panicles. Australia. (*Xanthorrhoea*). . . . . **Xanthorrhoeaceae**

75. Leaves in a single pair or in a whorl, reticulately veined. Tepals (4 –)6 – 10(– 16), the inner ones sometimes filiform to strap-shaped

('staminodes', actually 'petals'). (*Trilliaceae*). .... **Liliaceae**
— Leaves and flowers different. .... **Liliaceae**
76. Stigmas not twisted. Seeds fusiform with subulate ends. Exo- and endotesta with a cavity in between.—Flowers terminal on a naked, radical peduncle in dense capitules with leaf-like bracts. Lowland tropics of N.E. S. America. .... **Thurniaceae**
— Stigmas usually twisted. Seeds sometimes fusiform, but ends not subulate. Testa without such a cavity.—Flowers usually in variously compound inflorescences, rarely in involucrate capitules, or solitary. Plants of temperate zones and altitudes. .... **Juncaceae**
77. (42). Flowers actinomorphic. Fertile stamens 3–6. 78
— Flowers zygomorphic, usually bisexual. Fertile stamens 1 or 2.— Ovules numerous. .... **Orchidaceae**
78. Leaves parallel-nerved or scale-like. Ovary either 1, with 1 style and a simple to 3-lobed stigma, or ovaries 3, connate at base only. ... 79
— Leaves reticulately nerved. Ovary 1. Stigmas 2 or 3.—Leaves petiolate, usually broad. 80
79. Saprophytes. Leaves scale-like. Ovaries 3, connate at base only. (*Petrosaviaceae*). .... **Liliaceae**
— Autotrophic plants. Leaves well-developed, parallel-nerved. Ovary 1; style 1. (*Aletroideae, Ophiopogonoideae*). .... **Liliaceae**
80. Climbers. Flowers 3-merous. Ovary 3-locular and ovules axillary, rarely 1-locular and ovules parietal (*Rajania*). 81
— Stem erect. Flowers 4-merous. Ovary 1-locular, ovules apical.— Flowers bisexual. Anthers inappendiculate. (*Croomiaceae: Sticho-neuron*). .... **Stemonaceae**
81. Flowers unisexual. Connective not apically appendiculate. Ovules 2 per locule. .... **Dioscoreaceae**
— Flowers bisexual. Connective apically appendiculate. Ovules many per locule. (*Stenomeridaceae*). .... **Dioscoreaceae**
82. (36). Perianth corolloid. 83
— Perianth differentiated into a calyx and a corolla. 136
83. Ovary superior or nearly so. 84
— Ovary inferior or hemi-inferior. 104
84. Ovary 1, rarely ovaries 3, connate at base, perianth-segments then 6 (*Liliaceae*). 85
— Ovaries 3–more, free, when 3 perianth-segments 1–3. 102
85. Perianth-segments 6 or 8, rarely less, subequal when 4. 86
— Perianth-segments 4, very unequal.—Flowers in simple or bracteate spikes. Stamen 1. Ovary 1- or 3-locular. Ovules numerous. Stigma 1, punctiform or capitate. Endosperm fleshy. .... **Philydraceae**
86. Leaves only very rarely terminated by tendrils, then ovules numerous per locule and stigma undivided or with 3 short branches,

stipular tendrils sometimes present. 87

— Plants climbing with tendrils terminating the leaves.—Flowers in panicles, actinomorphic. Anthers dehiscing apically. Ovary 3-locular. Ovule 1 per locule, laterally attached. Stigma 1. Styles 3, elongated. Fruit a drupe. Endosperm mealy. Embryo small. (*Flagellaria*). .... **Flagellariaceae**

87. Anthers dehiscing with 1 slit or pore. Aquatics or plants of marshes.—Inflorescences spatheate.......................... 88

— Anthers usually dehiscing with 2 longitudinal slits, if with 1 slit or pore, then plants not aquatic or from marshes and either ericoid undershrubs or ovules atropous or hemitropous................. 89

88. Flowers in capitules subsessile at the base of the leaves, actinomorphic. Anthers with a terminal pore. Ovary 3-locular. Ovule 1 per locule, erect, basal. Embryo minute, broad.—Fruit a capsule. (*Maschalocephalus*). .... **Rapateaceae**

— Flowers in racemes, usually zygomorphic. Anthers introrse. Ovary either 3-locular with numerous, axillary ovules, or 1-locular with 1 apical, pendulous ovule. Embryo relatively large, linear. Perianth tubular at base. Style 1. Stigma 1. .... **Pontederiaceae**

89. Style 1, stigmas 3, usually twisted. 90

— Style 1 and stigmas 1, or 2, or 3, then usually short and not spirally twisted, or styles 3–5, free or connate at base only. 91

90. Leaves with distinct, usually tubular sheaths, 2- or 3-stichous. Inflorescence cymose with leaf- or scale-like bracts.—Plants grass- or rush-like, terrestrial. Perianth dry. Stamens 6 or less, the outer persistent; anthers basifix. Ovules 3–more per locule. Stigmas filiform.

Juncaceae

— Leaves broadly sheathing, usually in a spiral, rarely distichous. Inflorescence racemose, bracts large, usually coloured.—Habit different, terrestrial or epiphytic. Flowers in spikes or racemes. Ovules many per locule.............................. **Bromeliaceae**

91. Ovules usually anatropous, when atropous either stem woody and ovules pendulous, or ovary 1-locular and tepals 4. Embryo surrounded by the fleshy to cartilaginous endosperm, or basal and partly free...................................... 92

— Ovules usually atropous. Embryo apical, not surrounded by the mealy endosperm.—Stem herbaceous, leafy, nodose. Flowers 3-merous, usually in cincinni and blue. Filaments usually hairy. Ovary 3-locular. Ovules ascending, usually few per locule. **Commelinaceae**

92. Tepals 6. Funicle glabrous..................................... 93

— Tepals 4. Funicle hairy.—Erect or climbing herbs. Leaves reticulately nerved. Ovary 1-locular. Ovules several, basal, atropous.

Stemonaceae

14

93. Stamens 6 or more, rarely less, but then either staminodes present, or flowers not in racemes nor in panicles and not woolly, more or less actinomorphic. 94
— Stamens 1–3, staminodes sometimes present and flowers in racemes or in panicles, stamens sometimes 6, then flowers more or less zygomorphic, woolly. . . . . **Haemodoraceae**
94. Inflorescence with 1–several spathas, terminal on a leafless, unbranched peduncle, usually umbelloid, rarely a spadix-like spike, or 1-flowered.—Ovules 2–more per locule. 95
— Inflorescence without spathas, often with scale- or leaf-like bracts, rarely umbelloid. 96
95. Leaves not distichous. Flowers in umbels, rarely in a spadix-like spike (*Milula*). Anthers dorsifix, introrse, usually 6, rarely 2, 3, or 13. Stigma simple or 3-lobed.—Introduced in Australia and Tasmania only. (*Alliaceae*). . . . . **Liliaceae**
— Leaves distichous. Flowers solitary. Anthers basifix, extrorse, 3. Stigmas 3, thick, recurved. Tasmania. (*Isophysis*). . . . . **Iridaceae**
96. Leaves well-developed, or with leaf-like phylloclades. Flowers not involucrate, bracteate capitules, if so plants woody and/or ovules 2– more per locule. 97
— Plants rush-like, leaves reduced to the sheaths. Flowers 1–3 in an involucrate, bracteate capitule.—Anthers basifix. Ovule 1 per locule. Mediterranean. (*Aphyllanthaceae*). . . . . **Liliaceae**
97. Phylloclades usually absent, when present flowers axillary and filaments free. 98
— Phylloclades leaf-like. Flowers small, in terminal racemes or on the phylloclades. Filaments connate into a tube; anthers sessile, extrorse.—Fruit a berry. (*Ruscaceae*). . . . . **Liliaceae**
98. Leaves not very thick and fleshy and fibrous. Flowers solitary or in moderately sized inflorescences. 99
— Leaves very thick, fleshy and fibrous. Flowers in large to enormous spikes, racemes, or panicles, rarely in moderately sized ones, then ovule 1 per locule and fruit a berry (*Sansevieria*). (*Agavaceae*). . . . . . . **Liliaceae**
99. Evergreen undershrubs. Flowers solitary. Tepals 6. Anthers 6, erect, basifix. Ovary 1-locular and ovules 3, basal, erect, or 3-locular and ovule 1 per locule. S-, W-Australia. (*Calectasiaceae*).

Xanthorrhoeaceae
— Plants otherwise. 100
100. Shrubs or undershrubs, erect or climbing. Leaves reticulately veined. Inflorescences usually several-flowered. Fruit a berry. . . . 101
— Plants otherwise again, back to. 75
101. Plants usually climbing with or without stipular tendrils. Flowers usually dioecious, in umbels, rarely bisexual and in racemes or in

15

panicles (*Ripogonum*). Anthers basifix. Styles 3–5, free, or connate at base. (*Smilacaceae*). .... **Liliaceae**
— Tendrils absent. Flowers bisexual, usually in cymes, rarely solitary. Anthers dorsifix. Style 1, filiform; stigma small. (*Philesiaceae*).
                                                                          Liliaceae
102. (84). Autotrophic aquatics or plants from marshes. Leaves green, radical. Ovaries 3–6. Ovules 2–many per ovary. 103
— Non-green saprophytes. Leaves scale-like, cauline, alternate. Ovaries numerous. Ovule 1 per ovary. .... **Triuridaceae**
103. Leaves petiolate. Flowers in 1–several spikes. Tepals 2, rarely 1 or 3. Ovaries 3.—Aquatics, leaves submerged or floating.
                                                                 Aponogetonaceae
— Leaves non-petiolate. Flowers in umbels. Tepals and ovaries 6.— Plants from marshes. Leaves erect, linear, distichous. (*Butomus*).
                                                                       Butomaceae
104. (83). Fertile stamens 1–3. 105
— Fertile stamens 4–more..................................... 115
105. Fertile stamens 1 or 2, very rarely 3 and then, as usual, partly adnate with the style. Flowers usually zygomorphic............. 106
— Fertile stamens 3, very rarely 2, but always free from the style. Flowers usually actinomorphic. 110
106. Leaves pinninerved, petiolate. Flowers asymmetric, rarely zygomorphic, then leaves ligulate. Staminode(s) petaloid. Ovules and seeds not minute. Endosperm present. 107
— Leaves parallel-nerved, usually sessile, non-ligulate. Flowers zygomorphic, rarely nearly actinomorphic. Staminodes absent, rarely minute. Ovules and seeds minute. Endosperm absent.......... 109
107. Leaves non-ligulate. Flowers asymmetric. Outer tepals usually free. Anther with 1 fertile and 1 petaloid theca.................... 108
— Leaves ligulate. Flowers zygomorphic. Outer tepals connate. Anther with 2 fertile thecae, connective enlarged. .... **Zingiberaceae**
108. Petiole callose below the blade. Ovule 1 per locule, basal. Embryo curved. .... **Marantaceae**
— Petiole not callose below the blade. Ovules many per locule, axillary. Embryo straight......................... **Cannaceae**
109. Flowers usually distinctly zygomorphic. Fertile stamens usually 1, adnate to the stylar column, rarely 2 and (sub-)sessile on this column (*Cypripedieae*). Pollen grains coherent into clusters, or connate into pollinia, exceptionally free. Ovary usually 1-locular with parietal placentation, rarely 3-locular with axillary placentation, then flowers very zygomorphic (*Cypripedieae*).............. **Orchidaceae**
— Flowers nearly actinomorphic, the dorsal, inner tepal slightly concave. Fertile stamens 2 or 3, connate, partly free from the style.

Pollen grains free, finely granular. Ovary 3-locular with axillary placentation. (*Apostasiaceae*) . . . . **Orchidaceae**

110. Stamens opposite to the outer perianth-segments. 111
— Stamens opposite to the inner perianth-segments.—Anthers introrse or latrorse, or with a terminal pore or short slit. 112

111. Autotrophous herbs with green, often distichous leaves. Style-branches 3, rarely 2 (*Diplarrhena*), often petaloid . . . . **Iridaceae**
— Saprophytic non-green herbs with alternate, scale-like leaves. Style 3-lobed. Stigmas flattened.—Rhizome thin. Flowers bluish, ca. 1 cm long. Madagascar . . . . **Geosiridaceae**

112. Anthers with an apical pore or longitudinal slit. Style simple. Ovules usually not very numerous.—Plants autotrophous with well-developed leaves. 113
— Anthers with transverse, latrorse slits, rarely with longitudinal, introrse ones (*Oxygyne*), then, as usual, plants saprophytic. Style 3-fid. Ovules very numerous.—Leaves usually scale-like, radical when well-developed. Filaments very short. (*Burmannieae*).
Burmanniaceae

113. Staminodes absent. Anthers introrse with longitudinal slits. Perianth persistent in fruit. 114
— Staminodes 3. Anthers with apical pores or short slits. Perianth deciduous.—Ovary 3-locular. Ovules numerous. (*Tecophilaea*, *Tecophilaeaceae*). . . . . **Haemodoraceae**

114. Ovules numerous per locule, only 1 locule fertile in fruit. Placenta not peltate. (*Pauridia*) . . . . **Hypoxidaceae**
— Ovules 1–6 per locule, all locules fertile. Placenta peltate.
Haemodoraceae

115. Fertile stamens 5, staminode 0 or 1. Inflorescence with large, coloured bracts.—Large, rhizomatous to tree-like plants. Leaves pinninerved, often tearing between the nerves. Ovary 3-locular.   116
— Fertile stamens 6–more, rarely 4. Flowers usually actinomorphic.
118

116. Leaves distichous. Flowers bisexual. Fruit dehiscent 117
— Leaves alternate. Flowers usually unisexual, monoecious. Fruit leathery, indehiscent, or a pulpy berry.—Five tepals connate, 1 free. Ovules numerous per locule, axillary. Aril absent. . . . . **Musaceae**

117. Five tepals connate into a boat-shaped structure, 1 free. Ovule 1 per locule, basal. Fruit dehiscing into 3 cocci. Aril absent. (*Heliconiaceae*). . . . . **Musaceae**
— Tepals free, or the inner 2 oblique, forming a large, sagitate structure, the third short, boat-shaped. Ovules numerous per locule, axillary. Fruit a woody, loculicid capsule. Aril present, fimbriate. (*Strelitziaceae*). . . . . **Musaceae**

17

118. Ovary 1-locular, sometimes incompletely so. 119
— Ovary 3-locular. 127
119. Terrestrials. Flowers nearly always bisexual. Placentas 1–3. Endo-
sperm present, in minute seeds inconspicuous. 120
— Aquatics. Flowers nearly always unisexual, spatheate. Placentas
usually 6–more. Endosperm absent . . . . **Hydrocharitaceae**
120. Saprophytic, non-green plants. Leaves scale-like.—Flowers solitary,
or in bracteate, cymose racemes, or in capitules. Style simple. Stig-
mas 3, short. 121
— Autotrophic plants. Leaves well-developed.—Style simple or 3-
winged, stigma capitate to 3-fid, or styles 3. 122
121. Flowers actinomorphic. Stamens adnate to the perianth. Anthers introrse. (*Thismiaceae*). . . . . **Burmanniaceae**
— Flowers zygomorphic. Stamens free. Anthers extrorse. . . **Corsiaceae**
122. Flowers in a spatheate capitule or umbel, sometimes solitary. Stigma
3-fid to -lobed, sometimes inconspicuously so. 123
— Flowers in a spike, or in a raceme, without spathas, sometimes with
bracts. Stigma 1, capitate, or 3, filiform. 125
123. Leaves rarely reticulately veined, then flowers white. Flowers never
blackish. Style more or less terete. 124
— Leaves reticulately veined. Flowers blackish. Style with 3, some-
times deeply incised wings.—Ovules numerous. . . . . **Taccaceae**
124. Leaves radical. Flowers in an umbel, with a corona. Ovules 2–few.
(*Calostemma, Hymenocallis*). **Amaryllidaceae**
— Leaves cauline. Flowers solitary or sub-capitate, corona absent.
Ovules numerous. (*Leontochir, Schickendantzia: Alstroemeriaceae*).
Amaryllidaceae
125. Acaulescent, hairy herbs, or a few cauline leaves present, plants not
climbing. Inflorescences axillary. Anthers introrse. 126
— Thorny, scandent shrubs with tendrils. Cymes leaf-opposed. Anthers
extrorse.—Leaves cauline. Stigma 1, capitate. Ovules numerous.
Fruit a berry. (*Petermanniaceae*, also included in *Smilacaceae*).
Liliaceae
126. Leaves plicate. Capitules basal on a naked peduncle. Stigmas 3,
filiform. Ovules numerous. Fruit a berry. (*Curculigo*). **Hypoxidaceae**
— Leaves not plicate. Stem with a few leaves. Inflorescence a panicle
of cincinni. Stigma 1, capitate. Ovules 3 or 6. Fruit dry, dehiscent
(?). (*Phlebocarya, Lanaria*, the latter also in *Liliaceae* or *Tecophilaeaceae*). . . . . **Haemodoraceae**
127. Locules with several–many ovules, rarely 1 or 2, but then anthers
dehiscing with longitudinal slits, and/or ovary inferior. 128
— Locules with 2 ascending ovules.—Leaves broad, main nerves
curved, lateral nerves numerous. Flowers in racemes or in panicles.

Tepals nearly completely free. Anthers longer than the filaments, dehiscing apically. Ovary hemi-inferior, 3-lobed. Stigma lobed. Ovules anatropous. Seed 1 per fruit. Embryo lateral to the endosperm. (*Cyanastrum*, *Tecophilaeaceae* or *Cyanastraceae*).

Haemodoraceae

128. Ovary inferior, rarely hemi-inferior, then ovules many per locule, *or* flowers neither in spikes nor in racemes, *or* style and stigma simple. Embryo usually surrounded by the endosperm.................. 129
— Ovary hemi-inferior. Embryo lateral to the endosperm.—Flowers in spikes or in racemes. Perianth persistent in fruit. Filaments short. Style 3-fid, or simple with a 3-lobed stigma. Ovules 2–several per locule. (*Aletris*, *Ophiopogon*, *Peliosanthes*)............... **Liliaceae**
129. Plants woody, at least at base, then densely covered by a coat of fibres or roots.—Leaves radical or in terminal tufts............. 130
— Stem without such a cover, herbaceous, sometimes with a woody rhizome. 131
130. Stem densely covered by a coat of fibres or roots. Flowers solitary. Placentas laminar, ± peltate....................... **Velloziaceae**
— Stem without a coat. Inflorescences large to enormous. Placentas not laminar, nor peltate. (*Agavaceae*). ....·..... **Amaryllidaceae**
131. Leaves solitary at the end of each branch with a fascicle of flowers at its base.—Roots wiry. Tepals persistent in fruit. Ovules 2 per locule, serial, pendulous, apotropous. Fruit a winged berry. (*Trichopodaceae*). .... **Dioscoreaceae**
— Leaves several, usually radical. Inflorescences different......... 132
132. Inflorescence an umbel or an irregular raceme, rarely 1-flowered, provided with more or less membraneous spathas, when 1-flowered, occasionally with 1 leaf-like spatha. 133
— Inflorescence a raceme, or a panicle, or a capitule, rarely 1-flowered, without spathas, with or without scale- or leaf-like bracts.
134
133. Bulbs absent, roots swollen. Leaves cauline, often twisted at base. (*Alstroemeria*, *Bomarea*: *Alstroemeriaceae*)......... **Amaryllidaceae**
— Bulbs present. Leaves radical, not twisted at base... **Amaryllidaceae**
134. Placentas thickly laminar to peltate. Fruit a longitudinally dehiscent capsule.. 135
— Placentas not thickly laminar, nor peltate. Fruit dehiscing by a circular suture, or by short, vertical, subapical slits.—Leaves plicate or conspicuously nerved. Flowers small, white or yellow.

Hypoxidaceae

135. Plants glabrous. Inflorescence a lax raceme, or 1-flowered. Perianth deciduous. Anthers with an apical pore, rarely with longitudinal slits. (*Tecophilaeaceae*). .... **Haemodoraceae**

19

— Plants pubescent, hairs often branched. Inflorescence compound with cincinnate branches. Perianth persistent in fruit. Anthers with longitudinal slits............................ **Haemodoraceae**
136. (82). Ovary superior or nearly so. 137
— Ovary inferior or hemi-inferior. 151
137. Ovary 1, 1–5-locular. 138
— Ovaries 3–more, free or connate at base only, rarely also below the single style, flowers then solitary and involucrate in a secund spatheate spike (*Rapateaceae*). 149
138. Ovary 1-locular.—Ovules numerous. 139
— Ovary 2–5-locular. 143
139. Leaves oblong to ovate. Stamens 6–12. Ovules anatropous. Endosperm fleshy or cartilaginous.................................. 140
— Leaves linear. Fertile stamens 3. Ovules atropous. Endosperm mealy.. 141
140. Rhizomatous herbs. Leaves herbaceous, in a single pair or whorl. (*Trilliaceae*). .... **Liliaceae**
— Evergreen shrubs. Leaves leathery, numerous, alternate. (*Philesia, Philesiaceae*)..................................... **Liliaceae**
141. Leaves radical, stem sometimes with a few scales; apex entire. Flowers in spikes or in capitules. Anthers with longitudinal slits. 142
— Leaves cauline, apex bidentate. Flowers solitary or in umbels. Anthers with an apical pore.—Sepals equal, 3. Petals free.
Mayacaceae
142. Sepals homomorphic, 2 or 3. Petals connate. Style with 3 basal appendages. (*Abolbodaceae*). .... **Xyridaceae**
— Sepals heteromorphic, 3. Petals free. Style without basal appendages. .... **Xyridaceae**
143. Stamens 1–6. Ovary 2- or 3-locular. Ovules usually atropous. Embryo remote from the hilum............................ 144
— Stamens 6–12. Ovary 3–5-locular. Ovules anatropous. Embryo close to the hilum. 146
144. Flowers bisexual, not minute, usually in cincinni. Stigma 1, simple, or obscurely 3-lobed. Ovules usually several per locule, axillary, ascending..................................... 145
— Flowers unisexual, minute, in involucrate capitules, rarely axillary. Stigmas 2–6. Ovule 1 per locule, subapical, pendulous.—Leaves usually narrow. .... **Eriocaulaceae**
145. Flowers racemose, in spikes or in racemes.—Non-succulent, glandular-pubescent herbs. Leaves linear. Petals free. Stamens 6; filaments glabrous. Ovary 3-locular. Ovules 2 per locule. Fruit a capsule. Aru Isl., Australia. (*Cartonemataceae*)................ **Commelinaceae**

20

— Flowers cymose, usually in cincinni.—Plants otherwise.

Commelinaceae

146. Leaves often thorny-dentate, stiff and leathery, if herbaceous in a single pair or whorl. Anthers with introrse to latrorse longitudinal slits. Stigmas 3. 147

— Leaves entire, not stiff and leathery, numerous. Anthers with 1, or 2, or 4 apical pores. Style 1; stigma punctiform.—Flowers each with an involucre of several bracts, in spatheate capitules or spikes. Calyx and corolla contort. . . . . **Rapateaceae**

147. Leaves usually thorny-dentate, stiff, leathery, parallel-nerved, numerous, not in a single pair or whorl. Sepals and petals 3. 148

— Leaves herbaceous, entire, reticulately nerved, in a single pair or whorl. Sepals and petals (2–)3–5(–8). (*Trilliaceae*). . . . . . **Liliaceae**

148. Bracts green or brownish, rarely white. Flowers dioecious. Petals dry. Ovule 1 per locule, basal. Endosperm cartilaginous. Australia to New Guinea. (*Lomandra*, cf. also *Liliaceae s.s.*).

Xanthorrhoeaceae

— Bracts usually brightly coloured. Flowers usually bisexual. Petals not dry. Ovules few to numerous per locule, axillary. Endosperm mealy. Tropical America, many cultivated, occasionally escaping elsewhere in the tropics. . . . . **Bromeliaceae**

149. (137). Anthers extrorse or with apical pore(s). Ovules 1 or 2, basal, rarely a few and axial on 1 placenta. 150

— Anthers introrse. Ovules many, covering the entire inner face of the carpels.—Flowers solitary or in umbels. Fruit a follicle. (*Limnochar-itaceae*, sometimes also in *Alismataceae*). . . . . **Butomaceae**

150. Flowers without an involucre, in bracteate panicles, thyrses, or umbels, rarely solitary. Carpels 6–many, free, rarely connate at base, each with 1 free style. Fruit dry, indehiscent. . . . **Alismataceae**

— Flowers each involucrate by several bracts in a secund, spatheate spike. Carpels 3, connate at base and below the single style. Fruit a capsule, only 1 locule fertile. (*Spathanthus*). . . . . **Rapateaceae**

151. (136). Fertile stamen 1.—Flowers zygomorphic or asymmetric. . . 152

— Fertile stamens 2–more. 155

152. Leaves usually petiolate and pinninerved. Stamen free from the style or nearly so. Staminode(s) large, usually petaloid. Ovules and seeds not minute. Endosperm present. 153

— Leaves usually sessile and parallel-nerved. Stamen completely adnate to the style or nearly so. Staminodes minute or absent. Endosperm absent. . . . . **Orchidaceae**

153. Flowers zygomorphic or asymmetric. Sepals 3, connate, or with a deep slit. Anther with 2 fertile thecae, connective often enlarged.— Leaves ligulate. Ovules numerous. Embryo straight. 154

— Flowers asymmetric. Sepals 3, usually free. Anther with 1 fertile and 1 petaloid theca. Back to.............................. 108

154. Leaves distichous. Sheaths open. Ovary apically often with erect, sometimes large glands, 1- or 3-locular.—Plants aromatic.

Zingiberaceae

— Leaves in a spiral or 4-stichous. Sheaths initially closed. Ovary apically with depressed, supra-septal glands, 2- or 3-locular.—Supra-terranean parts not aromatic. (*Costaceae*)............ **Zingiberaceae**

155. Flowers usually actinomorphic. Stamens 2–16, sometimes some staminodial, when 5 plants aquatic. 156

— Flowers zygomorphic. Stamens 5.—Terrestrial. Leaves petiolate, blade large, oblong or ovate, pinninerved, transversally veined. Flowers perigynous, orchidaceous. .... **Lowiaceae**

156. Aquatic herbs. Flowers usually unisexual, solitary, or in spatheate cymes. Stamens 2–16. Anthers extrorse or latrorse. Ovary 1-locular, sometimes incompletely 6–15-locular. Endosperm absent.

Hydrocharitaceae

— Terrestrial or epiphytic plants. Flowers bisexual; inflorescences otherwise. Stamens 6. Anthers introrse. Ovary 3-locular. Endo-sperm copious. 157

157. Plants not climbing, often epiphytic, stem usually not developed. Leaves usually radical, margins usually thorny, usually lepidote. Flowers in spikes, or in racemes, or in panicles, or in capitules, usually with coloured bracts. Fruit a berry, or dry and indehiscent (*Bromelioideae*), or a septicide capsule (*Pitcairnioideae*).

Bromeliaceae

— Plants climbing or erect with a well-developed stem. Leaves alternate, entire, glabrous. Flowers in umbels with green bracts. Fruit a loculicide capsule. (*Bomarea, Alstroemeriaceae*). . . **Amaryllidaceae**

DICOTYLEDONES

158. (16). Plant with chlorophyll, if hemi-parasitic haustorial organs indistinct and usually lacking in the herbarium.—Try in case of doubt this lead also, as parasites and saprophytes have been taken up in the main key as well. 159

— Plant parasitic or saprophytic, *either* lacking chlorophyll, *or* hemi-parasitic and attached above the ground (e.g. as an epiphyte) to its host by haustorial organs, *or* with distinct subterranean connections.

2103

159. Perianth *either* absent, *or* simple, *or* composed of a calyx and at least one *free* petal, i.e. at least at base, rarely connate or cohering

22

in the middle or at the apex. ('*Archichlamydeae*'). 160
— Perianth differentiated into a calyx and petals, which are *all connate* at least at base. ('*Metachlamydeae*' or '*Sympetalae*'). 1572
160. Perianth *either* absent, *or* present, but then consisting of 2–7 (sub-)equal segments not differentiated into a calyx and a corolla, *or* absent in the female and/or bisexual flowers and present in the male ones, the latter then exceptionally with a calyx and a corolla. ('*Apetalae*'). 161
— Perianth *either* differentiated into a calyx and a corolla, *or* (rarely) consisting of 8–more (sub-)equal segments not clearly differentiated into a calyx and a corolla. ('*Choripetalae*'). 548

APETALAE

161. Bisexual and/or female flowers without a perianth, sometimes with bracts. 162
— Bisexual and/or female flowers with a perianth. 232
162. Flowers unisexual. 163
— Flowers bisexual or polygamous. 214
163. Male flowers without a perianth. 164
— Male flowers with a perianth. 192
164. Style or sessile stigma per flower 1, or 2–more, then connate at base. 165
— Styles or sessile stigmas per lower 2–more, free to base. 177
165. Ovary 2–4-locular or nearly so. 166
— Ovary 1-locular. 167
166. Ovules 1 or 2 per locule, pendulous. 175
— Ovules 2 per locule, ascending.—Leaves individed. Male flowers in catkins, each with 1(–3) bracts. Female flowers solitary in an involucre of many bracts. Stigmas deeply bifid. Ovary incompletely 2- or 3-locular. Australia, New Caledonia, Fiji. . . . . **Balanopaceae**
167. Ovule 1. 168
— Ovules 2–more. 174
168. Ovule pendulous sometimes from the middle of the adaxial wall. 169
— Ovule erect. 171
169. Flowers in a spike or in a panicle. 170
— Flowers on a spreading or thickened common receptacle, the female ones immersed in it.—Style present. Ovule anatropous. . . . **Moraceae**
170. Leaves alternate. Stipules absent. Ovary superior. Stigma decurrent, crenulate with a median groove. Madagascar. . . . . **Didymelaceae**
— Leaves opposite. Stipules present. Ovary inferior. Stigma terminal, truncate. S.E. Asia to New Zealand. . . . . **Chloranthaceae**

— Ovules 4 – more.—Flowers in a spike or a catkin............... 185
184. Trees. Leaves well-developed. Flowers solitary, or the male ones fasciculate. Stamens 6 – 10........................ **Eucommiaceae**
— Parasitic herbs. Leaves absent or scale-like. Inflorescence spadix-like. Stamens 2. (*Lophophytoideae*)............... **Balanophoraceae**
185. Submerged aquatic herbs. Leaves radical. Stipules absent. Stamen 1. Seed glabrous. Endosperm present. .... **Hydrostachydaceae**
— Shrubs or trees. Leaves alternate. Stipules present. Stamens 2 – more. Seeds hairy. Endosperm absent. .... **Salicaceae**
186. Ovules numerous per locule.—Stem woody. Stipules present. 187
— Ovules 1 or 2 per locule. 189
187. Leaves terminally tufted or alternate. Flowers in capitules. Stamens 8 – numerous. Styles and locules of the ovary 2. 188
— Leaves opposite. Flowers in catkin-like spikes. Stamens 3 – 8. Ovary-locules and styles 3 or 4................... **Myrothamnaceae**
188. Male inflorescence a terminal raceme of globose staminal clusters, each at first enveloped by a large membraneous bract. Ovules horizontal. (*Altingiaceae*). .... **Hamamelidaceae**
— Stamens 8 – 10 in distinct flowers. Ovules pendulous. (*Chunia*).
Hamamelidaceae
189. Terrestrials. Leaves usually alternate. Styles more or less apical.— Stipules usually present. Stamens 1 – many, free or connate. Styles or stigmas and locules of the ovary 2 or 3(– many)............. 190
— Aquatics with submerged or floating, opposite leaves. Styles gynobasic.—Stipules absent. Stamen 1. Styles 2. Ovary 4-locular. Fruit a schizocarp. .... **Callitrichaceae**
190. Ovules pendulous.—Fruit usually a capsule. 191
— Ovules basal, ascending.—Male flowers in catkins. Female flowers solitary, involucrate. Ovary incompletely 2- or 3-locular. Stigmas deeply bifid. Fruit an acorn-like drupe. Australia, New Caledonia, Fiji. .... **Balanopaceae**
191. Embryo minute, apical in copious, oily, blue endosperm. Fruit a 1-seeded drupe.—Leaves usually glaucous beneath. Stipules absent. Stamens 6 – 12. Pistillode absent. Ovary incompletely 2-locular. Stigmas 2, recurved or coiled. .... **Daphniphyllaceae**
— Embryo about as large as the endosperm. Fruit usually a capsule.— Stipules usually present. Ovary 3 – more-locular, rarely 2-locular, then completely so. (incl. *Peraceae, Uapacaceae*)..... **Euphorbiaceae**
192. (163). Style absent, stigma(s) sessile, if 2 – more connate at base and ovary 1 per flower....................................... 193
— Styles 2 – more, free to base, rarely ovaries 2 – 5, free, each with 1 style. 207
193. Stigma 1, sometimes 3- or 4-lobed.—Ovary 1-locular and ovule 1,

25

rarely locule inconspicuous and ovules 1 or 3. (*Balanophoraceae*).

27

basal. Stamens 2, if more, stigmas 2 or more. 220
— Leaves opposite. Stamens 1 or 3, connate and adnate to the ovary
or pistillode.—Stigma 1. Ovule pendulous. . . . . **Chloranthaceae**
220. Stipules present, sometimes adnate to the petiole. Flowers usually in
leaf-opposed spikes.—Shrubs, climbers or small trees. Ovule basal,
erect. Fruit a berry. Endosperm present. . . . . **Piperaceae**
— Stipules absent. Spikes axillary and/or terminal. 221
221. Shrubs. Spikes axillary. Stigmas 2. Ovule basal, erect with an elongated, recurved micropylar tube resembling a funicle. Fruit a drupe.
Endosperm absent.—New Caledonia. (*Canacomyrica*).   **Myricaceae**
— Herbs or undershrubs. Spikes axillary and/or terminal. Stigma
simple. Ovule basal, erect, without such a micropylar tube. Fruit a
berry. Endosperm present. (*Peperomiaceae*). . . . . **Piperaceae**
222. Leaves radical, tri-partite or -foliolate. Flowers in spikes. Stamens
(6–)9(–12), anthers with valves. Ovule erect. (*Podophyllaceae*).
Berberidaceae
— Leaves cauline, entire, in whorls. Flowers axillary, solitary. Stamen
1, anther with longitudinal slits. Ovule pendulous.—Marsh-plants.
Hippuridaceae
223. Stipules absent. Stamen 1. Stigmas 2 or 3.—Stem woody. (*Lacistemataceae*). . . . . **Flacourtiaceae**
— Stipules present. Stamens 5–more. Stigma 1. . . . . **Leguminosae**
224. (214). Locules of the ovary or ovaries 5–more. Stamens 8–many.
Stem woody.—Stipules absent. 226
— Locules of the ovary or ovaries 1–4. Stamens 1–10, rarely more,
then stem herbaceous. 229
225. (Deleted.)
226. Flowers either axillary, solitary or in clusters, or in terminal cymes
or panicles. Stamens homomorphic. Ovaries superior. Fruits samara-
like, or follicular, or capsular. 227
— Flowers terminal, solitary. Inner stamens petaloid, forming a
pseudo-perianth. Fruit a berry.—Perianth deciduous as a calyptra at
anthesis, leaving a scar. New Guinea, E. Australia. . . **Eupomatiaceae**
227. Stamens many. Ovaries more or less free, 6–18. Ovules 1–3 or
many per carpel. Fruits follicular or samara-like. 228
— Stamens 8–11. Ovary 8–15-locular. Ovules 4 per locule. Fruit a
capsule. New Caledonia. . . . . **Paracryphiaceae**
228. Flowers in terminal cymose racemes. Bracteoles several per flower.
Carpels laterally coherent, sessile. Ovules many per carpel. Fruits
follicular. Formosa, Japan. **Trochodendraceae**
— Flowers in axillary clusters. Bracteoles absent. Carpels free, stipitate. Ovules 1–3 per carpel. Fruits samara-like. Assam to Japan.
(*Eupteleaceae*). . . . . **Trochodendraceae**

229. Ovule 1. Ovary 1. 230
— Ovules 6 – many.—Herbs. 231
230. Ovary 1-locular. Ovule basal. . . . . **Piperaceae**
— Ovary 2-locular. Ovule apical.—Woody plants. (*Distyliopsis*).

Hamamelidaceae
231. Terrestrials. Flowers in spikes. Ovules 6–24, parietal, atropous. Endosperm present. . . . . **Saururaceae**
— Torrential aquatics, moss-like. Flowers spatheate. Ovules very many, central, anatropous. Endosperm absent. . . . . . . **Podostemaceae**
232. (161). Ovary or ovaries superior or nearly so, sometimes surrounded by the receptacle, but not adnate to it. 233
— Ovary inferior or hemi-inferior. 460
233. Ovary 1, undivided, or lobed. 234
— Ovaries 2 – more, free, or connate at base and/or the apex. 425
234. Ovary 1-locular, sometimes incompletely more-locular. 235
— Ovary completely 2 – more-locular, or nearly so. 338
235. Ovule 1. 236
— Ovules 2 – more. 283
236. Ovule or its funicle basal or nearly so. 237
— Ovule or its funicle apical or distinctly parietal. 267
237. Ovule atropous or nearly so, very rarely (*Canacomyrica*) with an elongated, recurved micropylar tube resembling a funicle. 238
— Ovule anatropous or campylotropous. 244
238. Style 1 or absent. Stigma 1, sometimes penicillate. 239
— Styles 2 – 4, free or connate at base.—Stamens usually 6 – 9. 240
239. Bark inside without silky, tough fibres. Stamens 1 – 5.—Perianth entire, or segments 2 – 5. 241
— Bark inside with silky, tough fibres. Stamens 8. . . . . . . **Thymelaeaceae**
240. Stipules usually connate into a sheath (ochrea). Perianth-segments 3 – 6. Endosperm copious, mealy. . . . . **Polygonaceae**
— Stipules absent. Perianth absent, but several bracteoles present. Endosperm absent.—Flowers in a spike. Style short, stigmas 2, long. Fruit a drupe. . . . . **Myricaceae**
241. Stipules absent. Flowers in spikes, or in racemes, or in fascicles, involucre absent. Stigma sessile, cushion-shaped or 2 – 5-lobed. Testa absent.—Woody plants. Endosperm copious. 242
— Stipules present, rarely absent, then either flowers solitary or in involucrate glomerules. Stigma linear or penicillate. Testa present.—Perianth-segments at least in the female flowers completely connate.

243
242. Perianth divided down to the disk into 3 – 5 segments. . . **Santalaceae**
— Perianth 4- or 5-lobed, male flowers moreover with 4 or 5 petals.—Flowers in racemes. (*Gjellerupia*). . . . . **Opiliaceae**

29

243. Stem usually herbaceous. Latex absent. Stamens incurved in bud.—
Leaves undivided or lobed. .... **Urticaceae**
— Stem woody. Latex present, rarely watery. Stamens erect in bud.

Moraceae
244. Ovule anatropous. Embryo straight. 245
— Ovule campylotropous. Embryo curved...................... 257
245. Stigmas 2 or 3. 246
— Stigma 1. 249
246. Flowers bisexual or polygamous............................ 247
— Flowers dioecious.—Tepals 1–5, imbricate. Stamens 3–5. Fruit a
drupe. Endosperm absent. (*Pistaciaceae*)........... **Anacardiaceae**
247. Tepals rarely 2. Stamens 7 or less. 248
— Tepals 2. Stamens 8–more. (*Bocconia*)............... **Papaveraceae**
248. Tepals 2–5, imbricate. Stamens hypogynous or perigynous, epitepalous, as many as, rarely more than the tepals. Fruit dry. Endosperm
present.—Leaves opposite. (incl. *Illecebraceae*). ... **Caryophyllaceae**
— Tepals 4–7, valvate. Stamens perigynous, alternitepalous. Fruit a
drupe. Endosperm scanty or absent. .... **Rhamnaceae**
249. Stamens perigynous on the upper margin of a more or less concave
receptacle, or inserted on the perianth....................... 250
— Stamens hypogynous or flowers unisexual and stamens on a central
column... 251
250. Stipules absent.—Leaves alternate. 255
— Stipules present.—Tepals 5–10. Stamens 1–4, or numerous. Fruit
dry. .... **Rosaceae**
251. Leaves usually opposite. Anthers introrse or latrorse.—Fruit indehiscent. Stamens 5 or more................................. 254
— Leaves alternate. Anthers extrorse or latrorse. 252
252. Young inflorescence resembling a young fir-cone. Filaments epitepalous, free or slightly adnate to the perianth. Ovule without integuments.—Trees, or shrubs, or lianas..................... **Opiliaceae**
— Young inflorescences not as above. Filaments alternitepalous, free
or connate. Ovule with 2 integuments. 253
253. Trees. Filaments completely connate. Fruit fleshy, dehiscent.

Myristicaceae
— Shrubs or lianas. Filaments free or connate at base. Fruit a drupe or
a samara. (*Petiveriaceae* = tribe *Rivineae*). .... **Phytolaccaceae**
254. Leaves with translucent dots and/or lines, crystals absent (lens!).
Tepals 4 or 6, free, imbricate. Fruit fleshy. Endosperm absent.—
Stamens many. .... **Guttiferae**
— Leaves with raphids and/or cystoliths, without translucent dots or
lines (lens!). Perianth corolloid, 4- or 5-dentate, plicate or contort.
Fruit dry. Endosperm present..................... **Nyctaginaceae**

30

255. Plants with peltate scales, at least on undersurface leaves. Stamens alternitepalous, 4 or twice as many as the 4–8 tepals.—Stamens inserted on the upper margin of the receptacle. Fruit fleshy.

Elaeagnaceae

— Plants with simple hairs or glabrous, rarely with medifixed hairs. Stamens epitepalous, 4 or 5, as many as the tepals. 256

256. Flowers usually elongate (at least so in S.E. Asia). Stamens adnate to the perianth-segments. Ovary usually stipitate (at least so in S.E. Asia). Integuments 2. S. Hemisphere. . . . . **Proteaceae**

— Flowers urceolate or shortly-cylindric. Stamens free or slightly adnate to the base of the perianth-segments. Ovary sessile. Integuments absent. S.E. Asia. (*Cansjera, Lepionurus*). . . . . **Opiliaceae**

257. (244). Perianth-segments either imbricate, rarely reduced to 1 tepal, or absent in the male flowers, or valvate and then either free, or stamens perigynous. 258

— Perianth undivided or 3–5-lobed, valvate or plicate, persistent in fruit, usually surrounded by bracts. Stamens hypogynous.—Leaves usually opposite. Perianth corolloid. Stigma 1. Plants usually with raphids and/or cystoliths (lens!). . . . . **Nyctaginaceae**

258. Stamens as many as the tepals, alternitepalous, or more. 259

— Stamens as many as the tepals, epitepalous, or less, rarely more, then leaves opposite and flowers in fascicles or in cymes, and stigmas 2–more. 264

259. Leaves alternate. 260

— Leaves opposite or in whorls.—Flowers solitary, or in glomerules, or in cymose panicles. Stigma simple. Endosperm present. (*Adenogramma*). . . . . **Aizoaceae**

260. Endosperm present.—Herbs, shrubs, or trees. Stigmas 1–5. 261

— Endosperm absent.—Shrubs, trees, or lianas. Flowers in fascicles or in panicles. Stigma 1. . . . . **Sapindaceae**

261. Flowers in spikes, or in racemes, or in panicles. Stigmas 1–5. . . 262

— Flowers in racemes of fascicles. Style 3-partite.—Stipules connate into a sheath (ochrea). . . . . **Polygonaceae**

262. Leaves simple. Flowers usually actinomorphic. Stamens free or connate at base only. 263

— Leaves compound. Flowers zygomorphic. Stamens 6, connate into 2 bundles of 3. (*Fumariaceae*). . . . . **Papaveraceae**

263. Anthers dorsifix. Fruit a drupe or a samara.—Flowers bisexual to monoecious. (*Petiveriaceae* = tribe *Rivineae*). . . . . **Phytolaccaceae**

— Anthers basifix. Fruit a berry.—Flowers dioecious. Stigmas 2. (*Achatocarpaceae*). . . . . **Phytolaccaceae**

264. Stipules absent. Tepals imbricate or valvate, rarely 1, or absent in the male flowers. Stamens as many as the tepals, or less, hypogy-

nous or nearly so, rarely distinctly perigynous, then either style simple, at least at base, or leaves alternate. 265
— Stipules present, rarely absent, then either stamens more than the tepals, or stamens distinctly perigynous and styles 2, free or partly connate, and leaves opposite. (incl. *Illecebraceae*). . **Caryophyllaceae**
265. Bracteoles present. Perianth more or less membranous or papyraceous. Filaments usually connate. Endosperm present. Embryo more or less curved. . . . . **Amaranthaceae**
— Bracteoles absent in bisexual and male flowers, rarely present, then embryo usually spirally curved. Perianth more or less herbaceous or membranous. Filaments free or nearly so.—Styles or stigmas or stigmatic lobes 2–5. 266
266. Tepals valvate.—Leaves alternate. Stipules absent. Tepals spongious in fruit. Stamens hypogynous. Embryo only slightly curved. Australia. . . . . **Dysphaniaceae**
— Tepals imbricate. . . . . **Chenopodiaceae**
267. (236). Stipules present. 268
— Stipules absent. 272
268. Fruit indehiscent.—Styles 1 or 2. 269
— Fruit dehiscent.—Leaves simple or lobed. Flowers unisexual, solitary, or in fascicles, or in spikes, or in racemes, or in panicles. Stamens hypogynous. Ovule with a caruncle. Endosperm present. Embryo straight. . . . . **Euphorbiaceae**
269. Stigmas 2–4, rarely 1, then flowers unisexual and all or the male flowers in a cymose, usually spike-like or capitate inflorescence or on a broadened common receptacle. 270
— Stigma 1. Flowers unisexual and solitary or bisexual.—Leaves usually compound. Endosperm absent. Embryo straight. . . **Rosaceae**
270. Flowers unisexual, the male ones in spike-, or in raceme-, or in capitule-like, or in paniculate inflorescences, or on a broadened common receptacle, rarely in cymose inflorescences, then stamens incurved in bud. 271
— Flowers bisexual or unisexual, then the male ones in lax cymes or in fascicles. Stamens straight in bud.—Shrubs or trees. Leaves simple, usually alternate (*Lozanella*: opposite). Stigmas 2–4. . . . . . **Ulmaceae**
271. Stipules free. Male flowers with 5 tepals and 5 stamens; female flowers with 1 tepal, enveloping the ovary. Filaments straight in bud.—Young leaves involute. . . . . **Cannabidaceae**
— Stipules connate, leaving an amplexicaul scar, if free leaves folded and filaments bent in bud. Flowers usually with 4 perianth-segments; stamens usually 4. . . . . **Moraceae**
272. Perianth present in all flowers. 273
— Perianth absent in male flowers, connate in the female ones.—

herbaceous. Perianth-tube globular to tubular. Stamens 8–10.
(*Galenia*). .... **Aizoaceae**

sent. Aril absent.—Perianth apert or imbricate. Stigma 1. . **Rosaceae**
296. Tepals 5. Stamens 5 or 10. 297
— Tepals 4 or 6. Stamens 1 or 4.—Fruit not dehiscing transversally. Embryo straight. 298
297. Herbs or undershrubs. Stigmas 2–4, sessile, apical. Fruit with a transverse suture. Endosperm copious. Embryo curved. (*Celosia*).

Amaranthaceae

— Shrubs. Style gynobasic; stigma 1, peltate. Fruit a nut or a drupe. Endosperm scanty. Embryo straight. S.W. Australia. (also in *Chrysobalanaceae* or *Rosaceae*). . . . . **Stylobasiaceae**
298. Tepals 4. Disk absent. Stamens 4. Placenta 1. 299
— Tepals 6. Extra-staminal disk present. Stamen 1. Placentas 2 (or 3). (*Lacistemataceae*). . . . . **Flacourtiaceae**
299. Stamens epitepalous. Endosperm absent. 300
— Stamens alternitepalous. Endosperm present.—Stamens 4. Fruit a drupe. (*Pyrenacantha*). . . . . **Icacinaceae**
300. Leaves not translucent-glandular punctate. Stamens usually adnate to the tepals, epigynous, rarely free and hypogynous (*Bellendena*, Tasmania). Style and stigma 1. . . . . **Proteaceae**
— Leaves translucent-glandular punctate. Stamens free from the tepals, hypogynous. Stigmas sessile, 2. S. Africa. (*Empleurum*).

Rutaceae

301. (283). Ovules basal or central, or laterally attached to the ovary-wall and subbasal in 2 rows. 302
— Ovules parietal, or laterally attached to the ovary-wall and then sometimes subapical and in 1 or 2 rows. 316
302. Stigma 1.—Fruit a capsule or a follicle. 303
— Stigmas 2–5. 309
303. Terrestrial plants. 304
— Aquatic, torrential herbs.—Leaves alternate. Perianth of 2 or 3 scales, apert. Stamens hypogynous. Endosperm absent. Embryo straight. . . . . **Podostemaceae**
304. Leaves alternate. 305
— Leaves opposite.—Perianth 5–more-merous, valvate or imbricate.

307

305. Flowers bisexual. Bracts, if any, not tubular. 306
— Flowers unisexual, dioecious. Bracts tubular.—N.W. Borneo.

Scyphostegiaceae

306. Perianth 4-merous, valvate. . . . . **Proteaceae**
— Perianth 4- or 5-merous, imbricate, usually dry and chaffy.— Flowers in spikes, or in racemes, or in panicles. Stamens usually connate at base. (*Celosieae*). **Amaranthaceae**
307. Perianth imbricate. Endosperm copious.—Perianth 5-partite. . . . 308

— Perianth valvate. Endosperm absent.—Stamens perigynous. Embryo straight.. **Lythraceae**
308. Stamens perigynous. Capsules dehiscing with a lid. Placentas axillary. Embryo curved. (*Trianthema*)................... **Aizoaceae**
— Stamens hypogynous. Capsule dehiscing with valves. Placenta central. Embryo straight. (*Glaux*)...................... **Primulaceae**
309. Stem woody.—Embryo straight. 310
— Stem herbaceous, sometimes woody at base, rarely entirely woody, but then embryo curved (*Amaranthaceae: Deeringia*)........... 312
310. Leaves opposite.—Perianth 5-partite, imbricate. Stamens 5, hypogynous, epitepalous.'.................... **Celastraceae**
— Leaves alternate. 311
311. Perianth 4–7-lobed, valvate. Stamens 4–7, perigynous, alternitepalous.. **Rhamnaceae**
— Tepals 5, imbricate. Stamens about 12, hypogynous.—Ovules 6 on 3 placentas. Fruit 3-winged. Mexico, Guatemala. (*Neopringlea*, of uncertain position, probably not belonging to:). .... **Flacourtiaceae**
312. Terrestrials. Perianth 4–6-partite, imbricate. Endosperm present. Embryo curved. 313
— Aquatic, torrential herbs. Perianth of 2 or 3 scales, apert. Endosperm absent. Embryo straight.—Leaves alternate. Stamens hypogynous, 1–3.......................... **Podostemaceae**
313. Leaves alternate.—Stipules absent. Stamens epitepalous....... 314
— Leaves opposite... 315
314. Lax herbs or undershrubs. Leaves well-developed, distant. Flowers in spikes, or in racemes, or in panicles. Filaments usually connate at base, 5 or more (*Celosieae*). .... **Amaranthaceae**
— Densely cushion-forming perennials. Leaves small, densely imbricate. Flowers solitary. Filaments free at base, usually 3. Kerguelen Isl. (*Lyallia*)............................... **Hectorellaceae**
315. Stamens hypogynous, rarely perigynous, then 4–more. Placenta central.—Perianth 4–6-partite. .... **Caryophyllaceae**
— Stamens perigynous, 1–3. Placenta basal.—Perianth 5-partite. Styles 2. (*Cypselea*). .... **Aizoaceae**
316. (301). Placenta 1, or ovules laterally attached to the ovary-wall in 1 row.................................... 317
— Placentas 2–more, or ovules laterally attached to the ovary-wall in 2 rows. 322
317. Leaves undivided, dentate or crenate.—Trees or shrubs. 318
— Leaves usually compound, rarely unifoliolate, or reduced to a leaf-like petiole, or digitately lobed or -sect, exceptionally simple, then plant herbaceous, leaves palmatinerved, stamens many (*Beesia*).—Fruit dry or a berry.................................... 320

36

318. Stamens many. 319
— Stamens 4, 3 staminodial.—Ovules laterally attached to the ovary-wall in 1 row. (*Placospermum*). . . . . **Proteaceae**
319. Ovules parietal, exceptionally subbasal. . . . . **Flacourtiaceae**
— Ovules apical.—S. America. (*Peridiscus, Whittonia*, also included in *Flacourtiaceae*). . . . . **Peridiscaceae**
320. Leaves digitately or pedately nerved, usually lobed or -sect. Stipules absent, petioles often sheathing. Stamens many, hypogynous. Nectaries, when present, between the stamens and the tepals.—Herbs.
321
— Leaves pinninerved, usually compound, rarely unifoliolate, or reduced to a leaf-like petiole. Stipules usually present. Stamens usually more or less perigynous. Nectaries, when present, between the stamens and the ovary. . . . . **Leguminosae**
321. Leaves 2, cauline. Carpels dehiscent along the ventral and dorsal sutures.—Rhizomatous herbs. Flowers solitary. Nectaries absent. Tepals 4. Japan. (*Glaucidiaceae*). . . . . **Ranunculaceae**
— Leaves several, usually at least some basal. Carpels dehiscent along the ventral suture or a berry.—Flowers usually in inflorescences. Nectaries present. (*Helleboraceae*). . . . . **Ranunculaceae**
322. Ovary sessile, when stipitate stigmas 2–6. 323
— Ovary stipitate.—Leaves alternate. Tepals 4. Stigma 1. Ovules campylotropous. Endosperm absent. Embryo strongly curved or involute. . . . . **Capparaceae**
323. Ovary initially apically open, closed after pollination.—Tepals 5 or 6. Stamens 10–30. Stigmas 3, sessile. (*Ochradenus*). . . . . **Resedaceae**
— Ovary completely closed. 324
324. Stamens 2–more. 325
— Stamen 1.—Shrubs or trees. Stipules absent. Flowers bisexual, in spikes. Style 1. Stigmas 2 or 3. Ovules few. C. America, West Indies. (*Lacistemataceae: Lacistema*). . . . . **Flacourtiaceae**
325. Perianth well-developed. 326
— Perianth actually absent, replaced by either a cupular, lobed disk, or 1 or 2 scales, which resemble a perianth.—Woody plants. Flowers dioecious, in spikes, or in racemes, or in catkins. Stigmas 2–4. Endosperm absent. . . . . **Salicaceae**
326. Stamens as many as the perianth-segments. 327
— Stamens more than the perianth-segments, rarely as many, then style and stigma 1.—Ovary sessile, rarely shortly stipitate then stamens numerous. 328
327. Stamens 4. Style 1. . . . . **Proteaceae**
— Stamens 5 or 6, rarely 8 or 9. Styles 3.—Stem usually climbing. Leaves alternate. Stipules present. Flowers bisexual, solitary or in

37

cymes. Perianth-segments 5 or 6, imbricate. Stamens perigynous. Styles free, or connate at base. Ovaries stipitate, rarely sessile, then stem herbaceous or woody at base only. (*Passiflora, Tryphostemma*). . . . . **Passifloraceae**

328. Plants autotrophic. Leaves well-developed. 329
— Herbaceous root-parasites. Leaves scale-like.—Perianth undivided or 4-lobed. Stamens numerous, connate. Style 1. Stigma undivided.
Rafflesiaceae

329. Stem herbaceous. 330
— Stem woody. 333

330. Leaves lobed to compound, the upper cauline sometimes simple and dentate, unarmed. Tepals 2 or 4. Stamens either 6 or many. Style 1 or stigma sessile. 331
— Leaves simple, serrate, underneath thorny on the nerves. Tepals 5 – 7. Stamens 10 – 14. Styles 2. (*Oresitrophe*). . . . . **Saxifragaceae**

331. Leaves lobed to dentate. Flowers actinomorphic. Stamens many, free. 332
— Leaves compound. Flowers zygomorphic. Stamens 6, connate into 2 bundles of 3.—Tepals 4, persistent during flowering. (*Fumariaceae*).
Papaveraceae

332. Flowers in panicles. Tepals 2, deciduous before flowering. (*Macleaya*). . . . . **Papaveraceae**
— Flowers solitary. Tepals 4, persistent during flowering. (*Glaucidiaceae*). . . . . **Ranunculaceae**

333. Perianth-segments 4 or more, rarely 3, imbricate, rarely valvate, but then tepals 3. (incl. *Neumanniaceae* and *Passifloraceae*: *Physena, Trichostephanus*). . . . . **Flacourtiaceae**
— Perianth-segments 3 – 8, valvate. 334

334. Style 1. Stigma 1, or 4 – 6. 335
— Styles 3 or 4, subulate with indistinct stigmas.—Trees. Leaves 3-plinerved. S. America. (*Flacourtiaceae*: *Peridiscus, Whittonia*).
Peridiscaceae

335. Stipules absent. Stamens perigynous.—Leaves opposite. 336
— Stipules present, but sometimes minute and early fugaceous. Stamens hypogynous. 337

336. Ovary incompletely 10 – 20-locular. Stamens numerous.—Fruit a berry. Endosperm absent. S.E. Asia. (*Sonneratia*). . . **Sonneratiaceae**
— Ovary apparently 2-locular with the septs touching each other. Stamens 5 (rarely 6). S. America. (*Crypteroniaceae*: *Alzatea*).
Lythraceae

337. Stamens on a cushion-shaped disk.—Leaves usually alternate, rarely opposite. Fruit a loculicide capsule, opening from the apex. (*Sloanea*) . . . . **Elaeocarpaceae**

— Disk absent.—Leaves alternate. Fruit a capsule, opening from the base and from the apex, fruitwall zigzag and intact. (*Itoa*).
Flacourtiaceae

350. Bark with silky, brownish fibres on the inside. Latex absent. Stipules absent. Ovary (3- or) 4-(or 5-)locular. Stigma 1. Ovule

without a caruncle. (*Aquilariaceae*: *Deltaria*, *Solmsia*).

Thymelaeaceae

— Bark without such fibres. Latex usually present. Stipules usually present. Ovary usually 3-locular. Stigmas 2–more. Ovule usually with a caruncle............................ **Euphorbiaceae**

351. Stipules absent. Flowers in spikes or in capitules. Fruit a capsule or a drupe. 352

— Stipules present. Flowers solitary or in glomerules. Fruit a berry. (*Doryalis*)..................................... **Flacourtiaceae**

352. Leaves opposite. Tepals (4–)5(–6). Ovary 3-locular. Styles 3. N. America. (*Simmondsiaceae*)......................... **Buxaceae**

— Leaves alternate to subverticillate. Tepals 4. Ovary 2-locular. Stigma 1, sessile. New Guinea, Australia. (also in *Aquifoliaceae*).

Sphenostemonaceae

353. Stamens hypogynous.............................. 354

— Stamens perigynous. 360

354. Leaves alternate, simple. 355

— Leaves opposite, tri-foliolate.—Herbs, woody at base. Stipules present. Flowers solitary. Perianth-segments 5, valvate. Stamens, styles 5. Embryo straight. (*Seetzenia*)............. **Zygophyllaceae**

355. Stipules absent. Stigmas 1 or 2. Embryo curved or ruminate. ... 356

— Stipules present. Stigmas 2–5, if 1 sessile and 3- or 4-lobed and ovary 3- or 4-locular. Embryo straight, not ruminate.—Woody plants. 357

356. Herbs, sometimes woody at base. Style 1, stigmas 1 or 2. Fruit capsular. Embryo curved........................ **Cruciferae**

— Woody plants. Stigma 1, sessile. Fruit a drupe. Embryo ruminate.— New Guinea, Australia. (also in *Aquifoliaceae*).

Sphenostemonaceae

357. Flowers solitary, or in fascicles, or in racemes. Stigmas 2–5. Fruit dry, indehiscent, or a berry, or a capsule.................... 358

— Flowers in thyrses. Stigma 1, sessile, discoid, lobed. Fruit a drupe.—Sumatra, Malaya. (*Endospermum*). .... **Euphorbiaceae**

358. Trees. Stigmas 2. Fruit dry, indehiscent or a drupe, 2-seeded.... 359

— Shrubs. Stigmas 3–5. Fruit a berry, 3–5-seeded.—Flowers solitary or in fascicles. Endosperm present. N.E. N. America. (*Nemopanthus*)...................................... **Aquifoliaceae**

359. Flowers in fascicles. Fruit indehiscent, winged. Endosperm absent. (*Ulmus*)....................................... **Ulmaceae**

— Flowers in racemes. Fruit a capsule. Endosperm present.

Hamamelidaceae

360. Bark inside without silky fibres. Stigmas 2–5................. 361

— Bark inside with tough, silky fibres. Stigma 1.—Shrubs or trees.

Stipules absent. Flowers in umbels or in capitules. Embryo straight.
Thymelaeaceae

361. Woody plants. Fruit dry, indehiscent. 362
— Herbs or undershrubs. Fruit a capsule.—Flowers solitary, or in glomerules, or in cymes. Stigmas 2–5. Embryo curved. (*Galenia, Plinthus*). .... **Aizoaceae**

362. Leaves opposite. Flowers solitary or in panicles. Stigmas 4. Embryo curved. Australia. (*Aphanopetalum*)................. **Cunoniaceae**
— Leaves alternate. Flowers in spikes, or in racemes, or in capitules. Stigmas and styles 2. Embryo straight. (*Hamamelioideae*).
Hamamelidaceae

363. (338). Ovules 2 per locule................................ 364
— Ovules 3–more per locule. 393

364. Ovules erect, or ascending, or patent, or one ascending and one descending. 365
— Ovules pendulous or descending. 375

365. Flowers bisexual. 366
— Flowers unisexual or polygamous. 372

366. Leaves usually opposite. Stamens perigynous................ 367
— Leaves usually alternate, rarely pseudo-verticillate then shrubs with 5 fertile stamens opening with pores. Stamens hypogynous...... 368

367. Herbs, at most woody at base. Stamens 5–many. Endosperm present..................................... **Aizoaceae**
— Ericoid shrubs. Stamens 4. Endosperm absent.—Flowers 4-merous, solitary or in spikes. (*Penaeae*). .... **Penaeaceae**

368. Stamens 8–more...................................... 369
— Stamens 5.—Flowers 5-merous, in cymes. Endosperm present. (*Lasiopetaleae*)................................. **Sterculiaceae**

369. Leaves without translucent glandular dots. Stipules present. Perianth-segments valvate.—Stamens 10–more............... 370
— Leaves with translucent glandular dots. Stipules absent. Perianth-segments imbricate.—Stamens 8–10. (*Asterolasia*)....... **Rutaceae**

370. Filaments free. 371
— Filaments connate. S. India, Ceylon. (*Cullenia*). ..... **Bombacaceae**

371. Leaves alternate or distichous. Anthers with longitudinal slits. Fruit not winged. (*Grewia*)................................. **Tiliaceae**
— Leaves opposite. Anthers with apical slits. Fruit with 3 wings. Burma, Thailand. (also in *Tiliaceae, Flacourtiaceae*). ... **Plagiopteraceae**

372. Leaves alternate. 373
— Leaves opposite.—Stigmas 2...................... **Aceraceae**

373. True perianth present. Disk present.—Leaves usually compound.374
— True perianth absent: male flowers (in catkins) with 1(–3) bracts, female flowers (solitary) involucrate. Disk absent.—Leaves simple.

Stamens (2–)5–6(–12), subsessile. Ovary incompletely 2- or 3-locular. Stigmas deeply bifid. Fruit an acorn-like drupe. Australia. New Caledonia, Fiji. .... **Balanopaceae**
374. Male flowers with a large, intra-staminal disk, lobed, between the lobes with 5 stamens and 5 staminodes. Female flowers with a 2- or 3-locular ovary, only 1 locule fertile. (*Alvaradoa*). ... **Simaroubaceae**
— Disk extra-staminal, small to well-developed. Female flowers with all the locules fertile. .... **Sapindaceae**
375. Ovules anatropous or campylotropous with a ventral raphe, or hemitropous. 376
— Ovules anatropous with a dorsal raphe, rarely atropous......... 389
376. Flowers bisexual. 377
— Flowers unisexual or polygamous. 381
377. Leaves alternate.—Herbs or undershrubs, if trees stipules present (sometimes early fugacious!)............................... 378
— Leaves opposite.—Stipules absent........................... 380
378. Herbs or undershrubs. Flowers not fascicled. Tepals free. Style developed. Fruit a silique, or dry and indehiscent, or a schizocarp.
379
— Trees. Flowers in fascicles. Perianth 3- or 4-(–6)-lobed or -partite. Fruit a drupe or a 3-valved capsule.—Stipules minute. Stamens 4. Stigmas 2 or 3. Endosperm present. Embryo straight. (*Aporosa, Drypetes*)...................................... **Euphorbiaceae**
379. Stipules absent. Flowers in racemes. Perianth-segments 4, imbricate. Stamens 2–6. Stigmas 1 or 2. Embryo curved. Endosperm scanty to absent. .... **Cruciferae**
— Stipules present. Flowers in cymes or panicles. Perianth-segments 5, valvate. Stamens 10–more. Stigmas 2–5. Endosperm present. (*Triumfetta*). .... **Tiliaceae**
380. Spiny shrubs. Flowers solitary or in fascicles. Perianth-segments 5, imbricate. Stamens 10. Stigmas 5. (*Rhynchotheca*, also in *Biebersteiniaceae, Ledocarpaceae, Vivianiaceae*). .... **Geraniaceae**
— Woody plants. Flowers in racemes or panicles. Perianth-segments 4, valvate. Stamens 2 or 3. .... **Oleaceae**
381. Ovary 2-locular or nearly so.—Shrubs or trees. Style 1 and stigmas 1 or 2, or stigmas 2, sessile........................... 382
— Ovary 3–more-locular, rarely 2-locular, then either styles 2, free, or connate at base only, or stigma 1, sessile..................... 386
382. Leaves paripinnate or trifoliolate....................... 383
— Leaves simple... 384
383. Leaves paripinnate. Stipules absent. Stamens 5–7. Tropical Africa, Asia and Australia. (*Ganophyllum*). .... **Sapindaceae**
— Leaves digitately trifoliolate. Stipules present, minute. Stamens

numerous. West Indies...................... **Picrodendraceae**

384. Stipules absent... 385
— Stipules present.—Leaves alternate. Flowers in cymose panicles. Stamens 10–18. Style and stigma 1. (*Heliocarpus*)........ **Tiliaceae**
385. Leaves opposite. Flowers in racemes or in panicles. Stamens (1–)2(–5). Ovary completely 2-locular. Style 1. Stigmas 1 or 2.......... **Oleaceae**
— Leaves alternate. Flowers in racemes. Stamens 6–12(–18?). Ovary incompletely 2-locular. Stigmas 2, sessile, recurved.—Embryo minute, apical, 4–6 times smaller than the copious endosperm. S.E. Asia...................................... **Daphniphyllaceae**
386. Leaves simple.. 387
— Leaves 3–7-foliolate. (*Oldfieldia*, *Piranhea*, and *Bischofiaceae*).
Euphorbiaceae
387. Male flowers in axillary triads of catkins. Female flowers axillary, solitary.—Leaves opposite. Stipules leathery, intrapetiolar. Disk absent. Stamens many. Ovary 3-locular. S.E. tropical Africa, Madagascar. (*Androstachydaceae*)..................... **Euphorbiaceae**
— Inflorescence and plants different. 388
388. Leaves alternate. Flowers in axillary catkin-like spikes or racemes. Tepals of male flowers imbricate. Disk absent in all flowers. Ovary 2-locular. Fruit a winged capsule. Endosperm scanty. Embryo large. Tropical Africa, S.E. Asia. (*Hymenocardiaceae*)..... **Euphorbiaceae**
— Plants different again. (incl. *Uapacaceae*). .... **Euphorbiaceae**
389. (375). Leaves alternate, simple. Flowers solitary or in fascicles. —Stipules present, often early caducous. Stamens 10–20. Styles 2–8. Endosperm present. Embryo straight. (*Doryalis*).
Flacourtiaceae
— Leaves opposite or alternate, but then flowers in spikes or in capitules. 390
390. Stipules absent. Inflorescences variously compound. Flowers unisexual or polygamous....................................... 391
— Stipules present. Flowers solitary, bisexual.—Perianth-segments 4. Stamens 8, perigynous. Styles 4................. **Geissolomataceae**
391. Fruit winged. Endosperm absent. Embryo curved....... **Aceraceae**
— Fruit not winged, sometimes horned. Endosperm present. Embryo straight.—Flowers in spikes or in capitules. 392
392. Leaves alternate or opposite. Stamens 4–6.............. **Buxaceae**
— Leaves opposite. Stamens many.—Locules of the ovary divided by secondary longitudinal septs. Colombia to Bolivia... **Stylocerataceae**
393. (363). Ovules basal, subbasal, parietal, or covering the septs nearly entirely. 394
— Ovules axillary, in 2-locular ovaries inserted on the middle of the sept.. 401

394. Styles 2–8.—Endosperm present.......................... 395
— Style 1... 396
395. Stamens many. Fruit indehiscent. (*Doryalis*)........ **Flacourtiaceae**
— Stamens 5 or 8. Fruit a capsule. (*Coelanthum, Macarthuria*).

                                                              Aizoaceae
396. Ovules more or less basal or on the septs..................... 397
— Ovules parietal on 2 placentas, connected by a false sept.—Herbs or undershrubs. Tepals 4. Stamens 1–6, hypogynous. Embryo curved.

                                                              Cruciferae
397. Leaves opposite. Stamens perigynous.—Perianth-segments valvate. Endosperm absent.. 398
— Leaves alternate. Stamens hypogynous........................ 400
398. Ericoid shrubs. Stipules present, very inconspicuous. Ovules basal, 4 in each of the 4 locules of the ovary. S. Africa. (*Penaeae*).

                                                        Penaeaceae
— Trees. Stipules absent or very inconspicuous. Ovules numerous. S.E. Asia, N. Australia....................................... 399
399. Flowers large, over 1 cm in diameter. Stamens numerous.

                                                      Sonneratiaceae
— Flowers small, 3 mm or less in diameter. Stamens 4 or 5. (*Crypteronia*, also in *Sonneratiaceae*).................. **Crypteroniaceae**
400. Leaves without beaker-shaped appendages. Flowers bisexual. Ovary stipitate. Fruit a berry or a drupe. Endosperm absent. Embryo curved...**Capparaceae**
— Leaves with beaker-shaped appendages. Plants dioecious. Ovary (sub-)sessile. Fruit a capsule. Endosperm present. Embryo straight.—Perianth-segments imbricate............... **Nepenthaceae**
401. Stamens hypogynous... 402
— Stamens perigynous or epigynous........................... 416
402. Perianth-segments valvate.................................... 403
— Perianth-segments imbricate or apert........................ 406
403. Woody plants, rarely undershrubs, then stigma 1 and embryo straight... 404
— Herbs or under shrubs. Stigmas several. Embryo curved. **Aizoaceae**
404. Flowers bisexual. Filaments free.—Stipules present. Stamens 8–many. Embryo straight.. 405
— Flowers unisexual, rarely bisexual, then fertile stamens 5. Flowers more or less connate, rarely free, then either styles several, or stipules absent.. **Sterculiaceae**
405. Stinging hairs absent. Leaves simple or compound, then opposite. Endosperm present... 407
— Stinging hairs present. Leaves pinnately compound, alternate. Endosperm absent.—Fruit a drupe with 2 pyrenes. N.E. Australia.

44

present. Flowers in racemes. Ovary 3–5-locular. (*Flacourtieae*).

Flacourtiaceae

420. Herbs or undershrubs.................................. 422
— Shrubs or trees.—Leaves opposite or in whorls. Perianth-segments valvate. Endosperm fleshy. Embryo straight................... 421
421. Stipules present. Stamens perigynous, 8–10.......... **Cunoniaceae**
— Stipules absent. Stamens epigynous, 5. (*Antoniaceae*: *Antonia*).

Loganiaceae

422. Endosperm fleshy. Embryo straight. 423
— Endosperm mealy. Embryo curved.—Flowers solitary, or in glomerules, or in cymes. .... **Aizoaceae**
423. Flowers in cymes or panicles. Stamens 5–10. 424
— Flowers solitary. Stamens 12.—Perianth-segments valvate. Stigmas 6. (*Asarum*). .... **Aristolochiaceae**
424. Flowers in cymes. Perianth-segments valvate. Styles 5–8, each with 1 capitate stigma.—Stamens 10. Carpels connate to half-way. E. Asia, E. N. America. (*Penthoraceae*, also in *Saxifragaceae*).

Crassulaceae

— Flowers in panicles. Perianth-segments imbricate. Stigmas 2 or 3.— Stamens 5–10. (*Saxifragoideae*). .... **Saxifragaceae**
425. (233). Ovule 1 per carpel, rarely accompanied by a second one, which is then early abortive. 426
— Ovules 2–more per carpel. 442
426. Stamens hypogynous.................................. 427
— Stamens perigynous. 434
427. Perianth-segments 2–6, rarely more, then either stamens more than perianth-segments, or flowers bisexual. 428
— Perianth-segments 6–more, stamens as many or less.—Woody plants. Flowers unisexual................................. 433
428. Leaves alternate, when opposite plants woody and leaves usually compound, endosperm not mealy, embryo minute, straight (*Clematis*).. 429
— Leaves opposite, simple.—Herbs. Stipules absent. Flowers in glomerules or cymes. Perianth-segments imbricate. Endosperm mealy. Embryo large, curved. (*Gisekia*). .... **Aizoaceae**
429. Stipules absent. Perianth-segments usually imbricate, if valvate plant annual, or a shrub, or a liana, and filaments free............... 430
— Stipules present. Perianth-segments valvate.—Trees. Flowers unisexual, in panicles. Filaments connate. (*Heritiera*, *Octolobus*).

Sterculiaceae

430. Fruit dry, if a drupe leaves compound and flowers in umbels. Endosperm fleshy, or cartilaginous, or horny. Embryo small to minute, straight... 431

— Fruit juicy. Leaves simple Endosperm mealy. Embryo large, curved.—Flowers in spikes or in racemes. .... **Phytolaccaceae**

431. Leaves with an open, dichotomous venation.—Himalaya to China.

432

— Leaves not so veined. (*Ranunculoideae*). .... **Ranunculaceae**

432. Tepals 2 (or 3), sepaloid. Stamens 2, rarely 1 or 3, staminodes absent. (*Circaeasteraceae*). .... **Ranunculaceae**

— Tepals 4–7, petaloid. Stamens 11–21, the outer staminodial. (*Kingdoniaceae*)........................... **Ranunculaceae**

433. Leaves usually simple, often 3- or 5-plinerved. Style usually gynobasic, rarely terminal. Fruit composed of drupelets with a distinctly sculptured endocarp. Pantropical................ **Menispermaceae**

— Leaves 3-foliolate (rarely simple just below the inflorescences). Style terminal. Fruit composed of berries. China....... **Sargentodoxaceae**

434. Leaves not tubular. 435

— Leaves tubular.—Herbs. Flowers in panicles. Flowers 6-merous. Stamens 12. Fruits follicular. Endosperm copious. ... **Cephalotaceae**

435. Stipules absent.—Leaves simple............................ 436

— Stipules present. 440

436. Woody plants. Leaves usually opposite. Fruit dry, indehiscent. Endosperm fleshy. Embryo straight. 437

— Herbs. Leaves alternate. Fruit a berry. Endosperm mealy. Embryo curved. .... **Phytolaccaceae**

437. Leaves opposite........................ 438

— Leaves alternate.—Anthers with introrse, longitudinal slits. Receptacle open. Carpels free, stipitate. New Caledonia. (also in *Monimiaceae*). .... **Amborellaceae**

438. Anthers with valves. 439

— Anthers with extrorse, longitudinal slits............. **Monimiaceae**

439. Flowers with an annular or flask-shaped disk (velum). Receptacle enclosing the carpels. Tropical Africa and America. (*Siparunaceae*).

Monimiaceae

— Flowers without a velum. Ovaries free. New Guinea, Australia to Chile. (*Atherospermataceae*)....................... **Monimiaceae**

440. Leaves alternate. Endosperm scanty or absent................ 441

— Leaves opposite or in whorls. Endosperm copious.—Woody plants. Flowers in panicles. Stamens 4–10. Carpels 2–5. New Guinea, Polynesia. (*Spiraeanthemum*). .... **Cunoniaceae**

441. Stipules extra-petiolarily connate. Perianth indistinct. Ovules atropous.—Trees. Leaves palmately lobed. Flowers unisexual, in capitules. Connective peltate. .... **Platanaceae**

— Stipules free, or adnate with the petiole. Perianth-segments 4 or 5. Ovules anatropous.—Anthers introrse. .... **Rosaceae**

47

ovules many per carpel (*Caltha*). Endosperm fleshy, or cartilaginous, or horny. 455

— Aquatics with peltate floating leaves and finely divided submerged ones. Endosperm mealy.—Flowers solitary. Tepals 6. Ovules 3 or 4 per carpel. (*Cabombaceae*). . . . . **Nymphaeaceae**

455. Leaves 2, cauline. Carpels dehiscent along the ventral and dorsal sutures.—Rhizomatous herb. Flowers solitary. Nectaries absent. Tepals 4. Japan. (*Glaucidiaceae*). . . . . **Ranunculaceae**

— Leaves several, usually at least a few basal. Carpels dehiscent along the ventral suture.—Flowers usually in inflorescences. Nectaries present. (*Helleboraceae*). . . . . **Ranunculaceae**

456. Leaves alternate. 457

— Leaves opposite.—Leaves undivided. Stipules absent. Flowers solitary, unisexual. Tepals (or bracteoles) 2–8, apert or imbricate. Filaments numerous, connate at base. . . . . **Cercidiphyllaceae**

457. Flowers unisexual or polygamous. 458

— Flowers bisexual.—Leaves undivided. Stipules absent. Flowers in catkin-like spikes. Tepals and stamens 4. (also in *Magnoliaceae*).

Tetracentraceae

458. Stipules present. 459

— Stipules absent.—Leaves compound. Flowers solitary or in racemes. Tepals 3 or 6. Stamens 6. Fruit juicy. . . . . **Lardizabalaceae**

459. Flowers solitary, or 2 or 3 together. Tepals 3, imbricate. Filaments free.—Leaves undivided. Stamens 6. Carpels 3. Fruit dry. Juan Fernandez. . . . . **Lactoridaceae**

— Flowers in panicles. Perianth-segments 3–5, connate at base, valvate. Filaments more or less connate. . . . . **Sterculiaceae**

460. (232). Ovary one, 1-locular, sometimes incompletely more-locular.

461

— Ovary completely 2–more-locular or nearly so, or ovaries several per flower. 509

461. Ovules quite distinct, at least in the older flowers. 467

— Ovules not clearly distinct from the ovary-tissue.—Parasites. 462

462. Fleshy, yellowish to brownish or red herbs without chlorophyl.

Balanophoraceae

— Parasitic shrubs with green leaves.—Male flowers without perianth, or perianth segments valvate; when stamens epipetalous, then as many as the segments. Stigma 1. 463

463. Flowers unisexual, the male flowers consisting of a group of up to 3 stamens. Fruit dry, with 3 feather-like bristles.—Epiphytic, shrubby, green parasites on *Nothofagus*. Temperate S. America.

Myzodendraceae

— Flowers bisexual or unisexual, in the latter case the male flowers either with a perianth, or (*Antidaphne*) consisting of a group of 4 stamens. Fruit usually fleshy, without feather-like bristles. 464

464. At least the bisexual or female flowers with a rim-like calyx (calyculus) below the corolla.—Flowers usually brightly coloured and usually bisexual, if flowers unisexual then plants dioecious.

Loranthaceae

— Calyx or calyculus absent.—Plants monoecious or dioecious. Flowers usually inconspicuous, greenish; unisexual. 465

465. Leaves usually decussate. Flowers in cymes or produced from the stem, not the leaf-axils (Tropical America, West Indies: *Dendrophthora, Phoradendron*). Anthers usually sessile or cohering.

Viscaceae

— Leaves usually alternate. Flowers in axillary or terminal racemose inflorescences. Anthers neither sessile, nor cohering. 466

466. Plants attached by means of large, distinct primary haustoria, sometimes also with secondary haustoria on creeping roots. Fruitwall without conspicuous longitudinal fibres. S. America, Mexico, Caribbean. . . . . **Eremolepidaceae**

— Plants without a distinct primary haustorium. Branches either leafy or with scales and then originating from endophytic parts. Fruitwall with conspicuous longitudinal fibres. S. E. Asia, New Guinea.

Santalaceae

467. Ovule 1. 468

— Ovules 2 – more. 487

468. Ovule more or less basal. 469

— Ovule more or less apical. 475

469. Ovule atropous.—Shrubs or trees. Flowers unisexual. Perianth calycoid. 470

— Ovule hemitropous or anatropous. 471

470. Leaves undivided. Stipules present. Flowers in glomerules. Stigma 1. Endosperm present. . . . . **Urticaceae**

— Leaves usually pinnately compound. Stipules absent. Flowers in spikes or in catkins. Stigmas 2. Endosperm absent. . . . **Juglandaceae**

471. Perianth calycoid. Stamens as many as the perianth-segments, epitepalous, or more, or less. Ovule hemitropous. Embryo curved.

472

— Perianth corolloid. Stamens as many as the perianth-segments, alternitepalous. Ovule usually anatropous. Embryo straight. 474

472. Perianth-segments imbricate. Stamens 1 – 5. Stigmas 2 – 5. Fruit a capsule or a nut. 473

— Perianth-segments valvate. Stamens 10 – 30. Stigma 1. Fruit a drupe.—Herbs. Leaves alternate. Stipules present. . . . **Theligonaceae**

473. Stipules absent.—Stamens 1 – 5. Stigmas 2 – 5. . . . . **Chenopodiaceae**

50

— Stipules present.—Herbs. Leaves opposite. Flowers in capitules, bisexual. Stamens 5. Stigmas 2. (*Paronychioideae*). **Caryophyllaceae**
474. Flowers solitary or in cymes, rarely in spikes or in racemes, or in capitules. Stigma 1, surrounded by a cup-shaped involucre. Endosperm present.—Flowers bisexual. . . . . **Goodeniaceae**
— Flowers usually in capitules. Stigmas 2, involucre absent. Endosperm absent. . . . . **Compositae**
475. Fleshy, herbaceous, red-brown root-parasites. Leaves scale-like.— Leaves alternate. Stipules absent. Flowers in a terminal, clavate spike. Stamen 1. Fruit a nut. **Cynomoriaceae**
— Autotrophic herbs or woody plants. Leaves well-developed. 476
476. Ovule anatropous or hemitropous. 477
— Ovule atropous.—Leaves opposite. Stipules present. Stamens 1–3, connate and adnate to the ovary. Stigma 1. Fruit a drupe. Endosperm present. . . . . **Chloranthaceae**
477. Stipules absent. 478
— Stipules present.—Flowers unisexual. Stamens 1–6. Ovule hemitropous. . . . . **Moraceae**
478. Leaves opposite, or in whorls, or radical, then sometimes spirally so.—Anthers with longitudinal slits. Ovule anatropous. Embryo straight. 479
— Leaves alternate, cauline. 481
479. Usually herbaceous terrestrials. Leaves opposite or radical. Perianth distinct. 480
— Aquatics. Leaves in whorls. Perianth an indistinct ridge.—Flowers solitary, bisexual. Stamen 1. Endosperm present. . . . . . **Hippuridaceae**
480. Leaves radical, spirally arranged. Tepals 2 or 3, calycoid, apert. Styles or stigmas 2. Endosperm fleshy.—Stamens 1 or 2. (*Gunneraceae*). . . . . **Haloragaceae**
— Leaves opposite, also when radical. Perianth-segments (3–)5, corolloid, imbricate. Style 1. Stigma 1, or 2- or 3-partite. Endosperm absent.—Stamens 1–4. . . . . **Valerianaceae**
481. Anthers with valves. Endosperm absent.—Woody plants. Perianth calycoid. Stigma 1. Embryo straight. 482
— Anthers not with valves. Endosperm present.—Leaves pinninerved.
484
482. Leaves tripli- or palmatinerved. Stamens in 1 whorl. Anthers 2-locular. 483
— Leaves pinninerved. Stamens in 3 whorls. Anthers 4-locular.— Flowers unisexual. Tepals and stamens 6. W. Africa. (*Hypodaphnis*). . . . . **Lauraceae**
483. Leaves without cystoliths. Tepals in 2 whorls, valvate. Stamens as many as the outer tepals. Cotyledons wrinkled. . . . . **Hernandiaceae**

— Leaves with cystoliths. Tepals in 1 whorl, imbricate. Stamens less than the tepals. Cotyledons plicate or convolute. (*Gyrocarpaceae*).

Hernandiaceae

484. Flowers bisexual, solitary or in fascicles, or in spikes. Ovule hemitropous. Embryo curved.—Herbs or undershrubs. Tepals 3–5. Endosperm mealy. (*Tetragoniaceae*). . . . . **Aizoaceae**

— Flowers arranged differently, if solitary flowers female and plants woody, dioecious. Ovule anatropous. Embryo straight. 485

485. Woody plants. Flowers unisexual, in racemes, or in panicles, or in capitules, rarely the male in umbels. Style 1, or 3, rarely 2. Fruit a drupe, or a berry, or samara-like. 486

— Woolly herbs. Flowers polygamous, in involucrate capituliform umbels, the outer flowers often male. Styles 2. Fruit woody, indehiscent.—Tepals 5, sepaloid, stamens 5, epitepalous. Endosperm cartiliginous. Australia, New Zealand. (*Hydrocotylaceae*: *Actinotus*).

Umbelliferae

486. Plants usually epiphytic or climbing, branches glabrous. Male flowers with calyx and corolla; stamens 5, alternitepalous. Female flowers in racemes or panicles; tepals sepaloid; style 1 with 3 stigmas, or style 3. Fruit a berry. (*Griselinaceae*). . . . . **Cornaceae**

— Plants terrestrial, branches silky-pubescent. Male flowers with corolloid tepals; stamens usually 10, diplostemonous. Female flowers solitary or in capitules, with calyx and corolla; style 1, rarely 2, each with 1 stigma. Fruit a drupe or samara-like. . . . . **Nyssaceae**

487. (467). Ovules 2. 488

— Ovules 3–more. 495

488. Stamens as many as the tepals, alternitepalous. Ovules ascending.

489

— Stamens as many as the perianth-segments, epitepalous, or more. Ovules pendulous. 490

489. Stipules present. Perianth calycoid. Stigma without an involucre. (*Condalia*). . . . . **Rhamnaceae**

— Stipules absent. Perianth corolloid. Stigma with a cup-shaped involucre. (*Scaevola*). . . . . **Goodeniaceae**

490. Stipules present, usually very distinct.—Flowers in spikes or in catkins. Male flowers without a perianth. Styles 2. Fruit dry. Endosperm absent. (incl. *Corylaceae*). **Betulaceae**

— Stipules absent. 491

491. Styles 2.—Leaves opposite. Flowers in spikes or in catkins. Male and female flowers with a perianth. Fruit juicy. Endosperm present. Sub(tropical) America. . . . . **Garryaceae**

— Style 1.—Flowers usually bisexual. 492

492. Leaves opposite.—Perianth-segments valvate. Endosperm present.

493

— Leaves alternate. 494

493. Stamens 3–6, as many as the perianth-segments. S.E. Asia, Australia.. **Santalaceae**

— Stamens 8, twice as many as the perianth-segments. S. Africa.

Grubbiaceae

494. Perianth-segments (3–)5(–8), valvate or imbricate. Stamens 10. Endosperm absent.............................. **Combretaceae**

— Perianth-segments 3–6, valvate, as many as the stamens. Endosperm present.................................... **Santalaceae**

495. Ovules 3–5.. 496

— Ovules 6–more... 501

496. Style 1. 497

— Styles 2–4, free.—Herbs or undershrubs. Tepals 3 or 4, stamens twice as many. Endosperm present. (*Haloragis, Laurembergia*).

Haloragaceae

497. Ovary incompletely 3-locular.—Woody plants. Flowers in spikes or in racemes. Perianth 4–6-lobed, corolloid. Stamens 4–6........ 498

— Ovary locular. 499

498. Flowers unisexual. Stigma 3–5-lobed, lobes bifid. (*Octoknemaceae*).

Olacaceae

— Flowers bisexual. Stigma shortly 3-lobed, lobes entire. (*Schoepfia*).

Olacaceae

499. Stamens 1–6. Ovules pendulous from a central, sometimes parietal placenta. Integuments and testa absent. Endosperm present..... 500

— Stamens 4–more, usually 8 or 10. Ovules apical, pendulous. Integuments and testa present. Endosperm absent.—Stigma 1.

Combretaceae

500. Male flowers without a perianth. Female flowers with 3 feathery appendages.—Epiphytic, shrubby, green parasites on *Nothofagus*. Flowers in spikes or in capitules. Temperate S. America.

Myzodendraceae

— Perianth present in all flowers, segments 3–6. Stigma 1, undivided or lobed.. **Santalaceae**

501. Ovary 1-locular, or incompletely more-locular with more than 2 ovules per 'locule'. 502

— Ovary, at least in the older flowers, incompletely 3–6-locular, ovules 2 per 'locule'.—Stem woody. Stipules present. Flowers unisexual or polygamous. Styles 3. Endosperm absent........ **Fagaceae**

502. Flowers unisexual or polygamous. 503

— Flowers bisexual. 505

503. Autotrophic, green plants. Leaves well-developed. Styles 2–more,

free. Endosperm absent...................................504
— Coloured, non-green parasites. Leaves scale-like. Style 1. Endosperm present.—Stamens numerous. ....**Rafflesiaceae**
504. Stipules absent. Placentas parietal. (incl. *Tetramelaceae*). **Datiscaceae**
— Stipules present. Placentas axillary. ....**Begoniaceae**
505. Autotrophic, green plants. Leaves well-developed. Stamens epitepalous, as many as the perianth-segments or more. 506
— Parasites, non-green, leafless. Stamens 3 or 4, as many as the tepals,
alternitepalous.—Flowers solitary. Tepals valvate. Style 1. Placentas
numerous. ....**Hydnoraceae**
506. Perianth 2–5-lobed, or undivided, nearly entire. 507
— Tepals 7 or 8.—Woody plants. Stamens numerous. Placentas 2 or 3.
(*Bembicia*). ....**Flacourtiaceae**
507. Perianth-segments 2 or 3, rarely 6, valvate, or perianth undivided,
nearly entire. Placentas 4–6. Styles connate into a column with
radiating stigmas. (*Aristolochia*). ....**Aristolochiaceae**
— Perianth-segments 4 or 5, imbricate. Placentas 2 or 3. Styles 2–4.—
Herbs...508
508. Staminodes absent. Stamens 4–10. Styles 2 or 3. (*Saxifragoideae*).
Saxifragaceae
— Staminodes 5. Stamens 5. Styles 3 or 4, short.—S. N. America,
Chile. (*Lepuropetalaceae*). ....**Saxifragaceae**
509. (460). Ovule 1 per locule...........................510
— Ovules 2–more per locule. 534
510. Ovule basal, subbasal, or median. 511
— Ovule apical or subapical. 515
511. Stigmas without a cupular involucre. 512
— Stigmas with a cupular involucre.—Stipules absent. Perianth corol-
loid. Stamens 5. Style 1. Stigmas 1 or 2. Fruit a drupe (*Scaevola*),
or dry, indehiscent (*Dampiera*). ....**Goodeniaceae**
512. Shrubs or trees. Perianth calycoid.....................513
— Herbs or undershrubs. Perianth corolloid.—Fruit dry, indehiscent.
514
513. Stamens 4 or 5. Disk intra-staminal. Ovary 2- or 3-locular. Style 1.
Rhamnaceae
— Stamens 2–many. Disk extra-staminal, annular to flask-shaped.
Ovary 4–many-locular. Styles 4–many. (*Siparunaceae*).
Monimiaceae
514. Leaves alternate. Stipules absent. Stamens numerous. Tepals free.—
Twining herbs. Style 1. Stigmas (3 or) 4. (*Agdestidaceae*).
Phytolaccaceae
— Leaves opposite or in whorls. Stipules present, sometimes leaf-like
(check axillary buds!). Tepals connate. Stamens 4 or 5.... **Rubiaceae**

55

— Herbs or undershrubs, usually aquatic.—Perianth 3- or 4-partite. Styles 3 or 4. . . . . **Haloragaceae**
527. Stipules present. 528
— Stipules absent. 529
528. Flowers in small epiphyllous fascicles from the midrib on the upper side of the leaf. Style 1, stigmas 3 or 4, recurved. Fruit a drupe. (*Helwingiaceae*, sometimes in *Araliaceae*). . . . . **Cornaceae**
— Flowers not epiphyllous. Styles 2. Fruit dry, indehiscent.

Hamamelidaceae

529. Flowers unisexual. Styles 3, or style very short and stigmas usually 3. 530
— Flowers bisexual. Style 1, stigma lobed. . . . . **Cornaceae**
530. Ovary 2-locular. Fruit a berry. New Zealand, S. America. (*Griseliniaceae*). . . . . **Cornaceae**
— Ovary 3- or 4-locular. Fruit a drupe. Himalaya, S. China. (*Torricelliaceae*). . . . . **Cornaceae**
531. Leaves radical and 2 cauline, opposite. Stamens 8–12, twice as many as the tepals.—Herbs. Flowers in a terminal glomerule. Styles 3–5. . . . . **Adoxaceae**
— Leaves alternate. Stamens as many as the tepals, rarely numerous.

532

532. Usually herbs. Flowers in umbels, rarely in capitules, or solitary. Disk 2-lobed or -partite. Styles 2. Fruit a schizocarp, or dry, indehiscent, very rarely a drupe and then flowers solitary. **Umbelliferae**
— Shrubs or trees, very rarely herbs (some *Araliaceae*) and then styles, as usual, 3–5. Flowers in umbels, or in capitules, or in racemes. Disk usually undivided. Fruit a drupe or a berry, very rarely a schizocarp then flowers in paniculate capitules (some *Araliaceae*). 533
533. Flowers in epiphyllous umbels or fascicles. Leaves simple, serrulate.—Himalaya, E. Asia. (*Helwingiaceae*, also included in *Araliaceae*). . . . . **Cornaceae**
— Flowers not epiphyllous. Leaves usually compound or divided, rarely simple then usually entire. . . . . **Araliaceae**
534. (509). Autotrophic plants with well-developed, green leaves. 535
— Parasites, non-green. Leaves absent or scale-like.—Stamens numerous. Style 1. Ovary with many locules. Flowers solitary. . . . . **Rafflesiaceae**
535. Perianth corolloid.—Leaves alternate. Ovules numerous per locule.

536

— Perianth calycoid. 539
536. Styles 1–3. Ovary 2- or 3-locular. 537
— Styles 4–6, or 1 with 4–6 stigmas. Ovary 4–6-locular.—Flowers bisexual. Perianth-segments connate, 1- or 2-labiate, or 3-lobed.

Stamens 5–more. Filaments short, thick, usually adnate to the style(s). Anthers extrorse or latrorse. Endosperm copious.

Aristolochiaceae

537. Herbs, undershrubs or shrubs. Flowers not connate into capitules. Anthers with slits or pores. Ovary inferior. Fruit a capsule or a berry. 538
— Trees. Flowers connate into capitules. Anthers with valves. Ovary hemi-inferior. Fruits connate into a syncarp. (*Exbucklandia*).

Hamamelidaceae

538. Flowers bisexual. Perianth 5-lobed with a dorsal slit. Stamens 5. Filaments connate. Style 1. Stigma 2-lobed. Endosperm present. (*Delissea*). .... **Campanulaceae**
— Flowers unisexual. Tepals 2 or 4 or perianth 2- or 4-lobed. Stamens many, free or an androphore. Styles 2 or 3, free or connate at base. Endosperm absent.................................. **Begoniaceae**

539. Plants herbaceous, at most woody at base.—Flowers bisexual. Ovules many per locule. 540
— Plants woody. 541

540. Perianth 4- or 5-partite, imbricate. Styles 2–5. Endosperm present.—Stamens 4–10. Fruit a capsule............ **Saxifragaceae**
— Tepals 3–5, free, valvate. Style 1, stigma undivided or lobed. Endosperm scanty to absent.—Stamens 1–8. Fruit a capsule or a berry. .... **Onagraceae**

541. Leaves opposite, if alternate glandular-punctate.—Flowers bisexual.
542
— Leaves alternate, not translucent-glandular punctate, rarely whorled (*Fagaceae*: *Trigonobalanus*). 544

542. Stipules present. Leaves not glandular-punctate. Perianth-segments valvate. Stamens 4–10. Endosperm present. 543
— Stipules absent, interpetiolary ridge sometimes present. Leaves translucent-glandular-punctate. Tepals free, imbricate or apert, or calyptrately connate. Stamens numerous. Endosperm absent.—Style 1, stigma undivided or lobed......................... **Myrtaceae**

543. Style 1. Stigma 8–10-lobed. Ovary 9–10-locular. Ovules many per locule. Fruit a berry.—Tepals 4 or 5. Stamens 8–10. (*Pellacalyx*).

Rhizophoraceae

— Styles 2, stigmas punctiform. Ovary 2-locular. Ovules 2–4 per locule. Fruit a nut.—Perianth (3- or) 4- or 5-partite to -fid. Stamens (6 or) 8 or 10. (*Ceratopetalum*, *Codia*)............... **Cunoniaceae**

544. Stipules present. Flowers small, unisexual or polygamous, solitary or in spikes, or in catkins, or in fascicles..................... 545
— Stipules absent. Flowers bisexual, fairly large, solitary, or in

57

racemes, or in panicles.—Perianth-segments valvate. Endosperm absent.. 547

545. Male flowers without a perianth. Anthers basifix. Ovary 2-locular. Ovules 6–many per locule. Fruit a capsule. Endosperm present. 546
— Male flowers with a perianth. Anthers dorsifix. Ovary 3–6-locular. Ovules 2 per locule. Fruit dry, indehiscent. Endosperm absent.

Fagaceae

546. Male inflorescence a terminal raceme of globose staminal clusters, each at first enveloped by a large membranous bract. Ovules horizontal. (*Altingiaceae*). .... **Hamamelidaceae**
— Stamens 10 in distinct flowers. Ovules pendulous. (*Exbucklandia*).

Hamamelidaceae

547. Flowers in racemes or in panicles. Tepals 3 or 4. Stamens 6 or 8. Styles 3 or 4. Ovary 3- or 4-locular. Ovules 2 per locule. Fruit a winged nut. (*Anisophylleaceae: Combretocarpus*). .. **Rhizophoraceae**
— Flowers solitary or in few-flowered cymes. Tepals 3–5. Stamens numerous. Style 1 with 4 short slender divaricate stigmas. Ovules 12 –20 per locule. Fruit a drupe. (*Foetidiaceae*)........ **Lecythidaceae**

CHORIPETALAE

548. (160). Ovary or ovaries **superior** on a small or dome-shaped receptacle. (When broadly sessile, try also the other lead).[1]......... 549
— Ovary **inferior**, or ovary or ovaries **hemi-inferior**, or **superior** on a distinctly enlarged, flat to hollow receptacle or hypanthium[2].... 1149
549. Disk absent, flowers occasionally with a corona................ 550
— Disk present, at least in the flowers of one sex. 926
550. Stamens 1–10. 551
— Stamens 11–more.. 805
551. Style 1 *per flower* (even when ovaries free), *either* simple with 1 or 2–more stigmas adjacent at base, *or* absent and stigma 1, sessile.

552

— Styles *either* 2–more *per flower*, free or connate at base but not up to the stigmas, *or* stigmas 2–more, sessile.................... 720

1 Thonner apparently sometimes interpreted petals and stamens as perigynous when the ovary is distinctly superior and the receptacle only slightly enlarged, e.g. some *Saxifragaceae*; in the 1917-version such plants were included after (the present) 1149. In this revision we have added such border-line cases also after 549 but perhaps not all instances have come to our attention.
2 Sometimes (e.g. *Rosaceae: Rubus*) the receptacle is flat or hollow, but also provided with a central dome.

565. Sepals 4..566
— Sepals 2.—Leaves divided. Corolla zygomorphic, more or less spurred. Stamens in 2 bundles, each with one 2-locular and two 1-locular anthers. Endosperm copious. Embryo basal, minute. (*Fumariaceae*)..............................**Papaveraceae**
566. Filaments all equal in length. (incl. *Cleomaceae*).......**Capparaceae**
— Filaments unequal in length.—Corolla actinomorphic, not spurred. Stamens free, or 4 pairwise connate, all anthers 2-locular. Endosperm scanty or absent. Embryo large, curved...........**Cruciferae**
567. (554). Anther with 1 transverse slit or with 1 pore. 568
— Anther with 2 longitudinal slits or 2 apical pores..............569
568. Flowers bisexual, zygomorphic. Stigma undivided or 2-partite. Embryo straight.—Stamens 8. ....**Polygalaceae**
— Flowers unisexual. Stigma 3–5-partite. Embryo curved.

Menispermaceae
569. Flowers actinomorphic. Stamens 4–9.—Trees or shrubs. Fruit a berry. 570
— Flowers zygomorphic, rarely actinomorphic, then stamens 10.

Leguminosae
570. Leaves alternate, usually compound. Stipules absent. Calyx 4- or 5-partite or -fid. Ovule pendulous......................**Meliaceae**
— Leaves opposite, undivided. Stipules or an intra-petiolary ridge present. Calyx 2–4-dentate. Ovule erect...........**Salvadoraceae**
571. (553). Ovules 2. 572
— Ovules 3–more..590
572. Ovules apical or central, pendulous. 573
— Ovules basal or lateral......................................574
573. Ovules apical. 586
— Ovules central.—Woody plants. Leaves undivided. Corolla actinomorphic, valvate. Stamens 5–10, free. Stigma 3-lobed.... **Olacaceae**
574. Ovules anatropous or hemitropous. 575
— Ovules atropous.—Shrubs or trees. Leaves usually compound. Stipules absent. Stamens 5 or 10, connate at base. Flowers actinomorphic. Ovules collateral.........................**Connaraceae**
575. Sepals 2, free. 576
— Sepals 3–more, rarely 2, then nearly completely connate.......578
576. Leaves divided. Stipules absent.—Herbs. Petals 4.............577
— Leaves undivided. Stipules present or leaves with axillary tufts of hairs.—Flowers actinomorphic. Stamens 1–5, free. ... **Portulacaceae**
577. Flowers zygomorphic; outer 1 or 2 petals saccate to spurred. Stamens 6 in 2 bundles. Stigma capitate. (*Fumariaceae*).

Papaveraceae
— Flowers actinomorphic; petals neither saccate, nor spurred. Stamens

4, free. Stigma 2-lobed. (*Pteridophyllaceae*). . . . . **Papaveraceae**
578. Stamens 6. 579
— Stamens 2–5, or 8–10. 580
579. Sepals and petals clearly differentiated, both 4. 566
— Perianth-segments numerous, not clearly differentiated into calyx and corolla.—Leaves bi- or ternately pinnately compound. China, Japan. (*Nandinaceae*). . . . . **Berberidaceae**
580. Stamens free, or 8–10, rarely 3–5 and filaments connate, then either leaves compound and/or flowers zygomorphic and/or stamens alternipetalous. 581
— Stamens 4–5, epipetalous. Filaments connate.—Leaves alternate, simple. Flowers actinomorphic. Calyx 5-fid, valvate. Petals contort.
Sterculiaceae
581. Leaves not translucent-glandular-punctate, or rarely so, then either stipules present, or flowers zygomorphic. 582
— Leaves translucent-glandular-punctate.—Shrubs or trees. Stipules absent. Flowers actinomorphic. . . . . **Rutaceae**
582. Ovules parietal. 583
— Ovules basal.—Leaves simple. 585
583. Placenta 1. 584
— Placentas 2.—Leaves alternate, simple. Anthers 5, nearly sessile, connate. Fruit a berry. . . . . **Violaceae**
584. Stipules usually present, sometimes early fugacious, when absent flowers actinomorphic. Anthers with longitudinal slits. 584a
— Stipules absent. Flowers zygomorphic. Anthers with 1 terminal pore.—Leaves usually densely hairy, simple, rarely 3-foliolate. Sepals 4 or 5, inbricate, free, unequal. Endosperm absent. America. . . . . **Krameriaceae**
584a. Stipules present, sometimes early fugacious. Flowers zygomorphic or actinomorphic. Calyx-segments and petals usually 5. Endosperm scanty or absent, rarely copious. . . . . **Leguminosae**
— Stipules absent. Flowers actinomorphic. Sepals 3. Petals 6. Endosperm ruminate.—Medullary rays in twigs on cross-section usually regular and distinct, dilating in the bark. Leaves simple, undivided. Sepals valvate. Petals imbricate. . . . . **Annonaceae**
585. Leaves opposite. Stamens 4 or 5. . . . . **Salvadoraceae**
— Leaves alternate. Stamens 10. (*Guilfoylia*). . . . . **Simaroubaceae**
586. (573). Petals 4. Stamens 6, free, or the 4 longer ones pairwise connate.—Herbs or undershrubs. Leaves simple. Stipules absent.
Cruciferae
— Stamens 2–5, or 7–10, rarely 6, then either all connate, or petals 3. 587
587. Filaments free, rarely connate, then either flowers zygomorphic or

stipules present. 588

— Stipules absent. Flowers actinomorphic. Filaments connate.—Shrubs or trees. Leaves pinnately compound. . . . . **Meliaceae**

588. Leaves not translucent-glandular-punctate, or rarely so, then either stipules present or flowers zygomorphic. 589

— Leaves translucent-glandular-punctate. Stipules absent. Flowers actinomorphic.—Shrubs or trees. Leaves compound. Calyx 3- or 4-dentate. Petals 3 or 4, imbricate. Stamens 6–8. Fruit a drupe.

Rutaceae

589. Leaves simple. Stipules absent. Flowers actinomorphic.—Shrubs or trees. Stamens 4 or 5, free. Fruit a drupe or a nut. . . . . . **Icacinaceae**

— Leaves compound, rarely simple, then either stipules present or flowers zygomorphic. . . . . **Leguminosae**

590. (571). Placenta 1, basal or central. 591

— Placentas 1–several, parietal. 598

591. Ovules erect or laterally attached on a central placenta. 592

— Ovules pendulous.—Stigma 1. Ovules 3–5. . . . . **Olacaceae**

592. Anthers with longitudinal slits, rarely with apical pores.—Leaves simple. 594

— Anthers with valves.—Leaves alternate or radical. Stamens 6. Stigma 1. Ovules basal. 593

593. Stem herbaceous, rhizome tuberous or creeping, fleshy.—Leaves radical or cauline. (*Leonticaceae*). . . . . **Berberidaceae**

— Stem woody. . . . . **Berberidaceae**

594. Sepals 3–more, or calyx 2- or 3-fid. 595

— Sepals 2.—Stigmas 2–8. . . . . **Portulacaceae**

595. Stem woody. Leaves alternate. Stigma undivided or lobed. 596

— Stem herbaceous, at most woody at base. Leaves opposite. Stigma grooved, lobed or divided.—Leaves undivided. Fruit a capsule.

Caryophyllaceae

596. Petals and stamens 4 or 5.—Fruit a drupe. . . . . **Myrsinaceae**

— Petals 5. Stamens 10. 597

597. Leaves simple. Calyx 2- or 3- fid. Style terminal. Tropical Africa. (*Afrostyrax*, formerly in *Styracaceae*). . . . . **Huaceae**

— Leaves pinnately compound. Sepals 5. Style gynobasic. Mexico. (*Recchia*). . . . . **Simaroubaceae**

598. Placentas 2–more. 599

— Placenta 1. 601

599. Petals 3. 600

— Petals 4–more. 606

600. Non-green parasites. Leaves scale-like. Sepals 3. Stamens 6. Embryo very small. (*Hypopitys*). . . . . **Monotropaceae**

— Autotrophic, green plants. Leaves well-developed. Sepals 5.

Stamens 3–10. Embryo large. (*Lechea*). .... **Cistaceae**

601. More stamens than petals, rarely as many or less, then either stem woody or stipules present. 602
— Stamens as many as the petals, epipetalous.—Herbs. Stipules absent. Flowers actinomorphic. 605

602. Stipules present, sometimes early fugacious. Flowers zygomorphic or actinomorphic.—Calyx-segments and petals usually 5. Endosperm scanty or absent, rarely copious. .... **Leguminosae**
— Stipules absent. Flowers actinomorphic. 603

603. Sepals and petals (4 or) 5, imbricate.—Twigs without such medullary rays as in *Annonaceae* (see sub 604). Leaves simple or compound, hairy or glabrous. Endosperm, if present, not ruminate.

Connaraceae

— Sepals either 3 and valvate or calyptrate and caducous, or persistent and then cup- or saucer-shaped, entire or ruptured into more or less irregular 'lobes'. 604

604. Wood with vessels. Twigs on cross-section with a regular pattern of radial medullary rays, dilating in the bark. Leaves hairy or glabrous. Calyx either with distinct lobes or sepals free. Endosperm ruminate.

Annonaceae

— Wood without vessels. Twigs without such medullary rays. Leaves glabrous. Calyx either calyptrate and caducous, or persistent, then cup- or saucer-shaped, entire or ruptured into more or less irregular 'lobes'. Endosperm not ruminate. (*Belliolum*, *Bubbia*, *Drymis*, *Pseudowintera*)................................... **Winteraceae**

605. Sepals 12–15, or stamens 4. (*Epimedium*, *Vancouveria*).

Berberidaceae

— Sepals 4–8, or stamens 6–more. (*Podophyllaceae*). . **Berberidaceae**

606. Sepals 2 or 3.. 607
— Sepals 4–more. 611

607. Petals 4 or 6... 608
— Petals 5 or 10... 610

608. Petals not spurred. Stamens 4. 609
— Outer 1 or 2 petals saccate to spurred. Stamens 6 in 2 bundles of 3. (*Fumariaceae*). .... **Papaveraceae**

609. Flowers actinomorphic, petals entire. Ovules 3 or 4. (*Pteridophyllaceae*). .... **Papaveraceae**
— Flowers more or less zygomorphic, outer petal 3-lobed, inner 3-partite. Ovules many. (*Hypecoaceae*)................ **Papaveraceae**

610. Stamens connate, 10. Fruit a berry.—Shrubs or trees. (*Canella*, *Warburgia*). .... **Canellaceae**
— Stamens free, 10–more. Fruit a capsule. (*Hudsonia*)...... **Cistaceae**

611. Sepals and petals 4. 612

— Sepals and petals 5–8. 616
612. Leaves alternate. Stigmas 1 or 2. 613
— Leaves opposite. Stigmas 2–4.—Herbs or undershrubs. Leaves simple, often ericoid. Stipules absent. Fruit a capsule. Salty areas.
Frankeniaceae
613. Leaves usually simple. Filaments free or connate at base only. . . 614
— Leaves usually pinnately compound. Filaments connate.—Woody plants. Stamens 8. Stigma 1. Fruit a capsule............. **Meliaceae**
614. Stipules absent. Ovules campylotropous. Fruit rarely a berry. Endosperm absent. Embryo curved. 615
— Stipules present. Ovules anatropous. Fruit a berry. Endosperm present. Embryo straight.—Plants woody. Flowers often before the leaves. Seeds arillate. Himalaya to Japan. .... **Stachyuraceae**
615. Stamens 6, 2 shorter than the others. Ovary usually sessile. Stigmas 1 or 2.—Herbs or undershrubs. Fruit dry, dehiscent...... **Cruciferae**
— Stamens either 6, equal, or 4, or 8. Ovary usually stipitate. Stigma 1... **Capparaceae**
616. Calyx imbricate or apert. 617
— Calyx valvate. 618
617. Anthers introrse, latrorse, or apically dehiscent.............. 619
— Anthers extrorse.—Insectivorous herbs. Leaves radical, glandular. Stipules present. Flowers bisexual. Stamens 5, filaments long. Staminodes absent..................................... **Droseraceae**
618. Leaves opposite. Stamens 4–6. Stigmas 2–4.—Halophilous herbs or undershrubs. Anthers extrorse or latrorse.......... **Frankeniaceae**
— Leaves alternate. Stamens 6–9. Stigmas 2.—Shrubs or small trees. E. Australia, Tasmania. (*Escalloniaceae: Anopterus*). **Saxifragaceae**
619. Stamens 5–8. 620
— Stamens 10. 621
620. Flowers bisexual or polygamous............................. 625
— Flowers unisexual.—Flowers actinomorphic. Stamens 5 or 6. Corona present. (*Adenia*). .... **Passifloraceae**
621. Leaves simple, undivided.................................. 622
— Leaves pinnately-compound.—Woody plants. Leaves translucent-glandular-punctate. Fruit a berry....................... **Rutaceae**
622. Herbs. Anthers adnate, usually with apical pores or slits, if with longitudinal slits, plants non-green saprophytes. 623
— Woody plants, rarely herbs. Anthers versatile, usually with longitudinal slits. 624
623. Autotrophic plants with well-developed, green leaves. Anthers incurved in bud, with 2 apical pores or tubules. .... **Pyrolaceae**
— Non-green saprophytes without well-developed leaves. Anthers

64

erect in bud, thecae with a common slit, or with 2 longitudinal slits.

Monotropaceae

624. Herbs or undershrubs, often ericoid.—Anthers introrse. Ovary completely 1-locular. Stigmas 3 or 4. Endosperm absent. (*Tamariceae*).. **Tamaricaceae**

— Woody plants, non-ericoid.—Stamens epipetalous. Fruit a capsule. (cf. *Homalium*). .... **Flacourtiaceae**

625. Fertile stamens 5–8, staminodes absent. Fruit a loculicide capsule, or a berry, or dry, indehiscent, 1-seeded. 626

— Fertile stamens 5, rarely 8; staminodes in an outer whorl. Fruit a septicide capsule.—Stipules present.................. **Ochnaceae**

626. Stipules absent. Flowers actinomorphic or nearly so.—Woody plants... 627

— Stipules present, rarely absent, then flowers distinctly zygomorphic. —Filaments short. Anthers usually appendiculate. .... **Violaceae**

627. Stamens 5, alternipetalous. Embryo minute. .... **Pittosporaceae**

— Stamens 5–8, epipetalous. Embryo relatively large.—Leaves alternate or in whorls. Anthers latrorse. Connective broad. (cf. *Gerrardina*).. **Flacourtiaceae**

628. (552). Ovule 1 per locule................................... 629

— Ovules 2–more per locule. 659

629. Ovule erect, ascending, or patent.—Stamens as many as the petals, or more... 630

— Ovule pendulous or descending. 638

630. Leaves opposite... 631

— Leaves alternate. 632

631. Stipules present. Flowers unisexual.—Flowers solitary, or in spikes, or in racemes, or in panicles. Petals 4. Stamens 4, free.

Salvadoraceae

— Stipules absent. Flowers polygamous.—Leaves with translucent to black glandular dots or lines.......................... **Guttiferae**

632. Flowers bisexual, at least apparently so, or polygamous........ 633

— Flowers unisexual.—Stem woody. Stipules present. Flowers in fascicles. Petals 4. Stamens 4, free................... **Celastraceae**

633. Stamens not 4. 634

— Stamens 4.—Trees. Petals 4. (*Tetrameristaceae*). .... **Theaceae**

634. Aquatics or marsh-plants.—Stamens and petals 5. (*Hydrocera*).

Balsaminaceae

— Terrestrial plants....................................... 635

635. Shrubs or trees. Stamens 3–5, or 7–10, rarely 6, then petals 3 or 6. 636

— Herbs or undershrubs. Petals 4. Stamens 6.............. **Cruciferae**

636. Stipules present. 637

— Stipules absent.—Leaves usually compound, rarely unifoliolate. Filaments connate. Anthers with longitudinal slits........ **Meliaceae**
637. Calyx imbricate. Filaments free.—Anthers usually with apical pores.
                                                                        Ochnaceae
— Calyx valvate. Filaments free or connate.—Stamens 5, epipetalous.
                                                                    Sterculiaceae
638. Flowers bisexual or polygamous............................. 641
— Flowers unisexual. 639
639. Petals valvate. Endosperm absent. (*Picrolemma*). ... **Simaroubaceae**
— Petals imbricate. Endosperm present.—Ovules usually with a caruncle... 640
640. Petals in a whorl. Micropyle pointing outward........ **Euphorbiaceae**
— Petals decussate (2 + 2, rarely only 2). Micropyle pointing inward.—Ovary 2-locular. Stigma 1, sessile. Seed ruminate. (also included in *Aquifoliaceae*)..................... **Sphenostemonaceae**
641. Filaments free, stamens rarely paired with connate filaments. ... 642
— Filaments all connate at least at base. 654
642. Anthers with 1 or 2 apical pores. 643
— Anthers with 2 longitudinal slits............................. 645
643. Stipules absent. Flowers solitary, axillary.—Shrubs............. 644
— Stipules present. Flowers in racemes or in panicles.—Calyx and corolla imbricate. Anthers with 2 apical pores. .... **Ochnaceae**
644. Leaves, young stems, and calyx with long, club-shaped glands. Calyx and corolla imbricate. Anthers with 2 apical pores or short slits.—Leaves alternate. S. Africa. .... **Roridulaceae**
— Plants without such glands. Calyx valvate, corolla induplicative-valvate. Anthers with 1 apical pore.—Leaves alternate, or opposite, or in whorls. Australia........................ **Tremandraceae**
645. Usually trees or shrubs. Petals 5, rarely 3 or 6, or 4, then either flowers polygamous or stamens 8 or stigma 1, sessile. 646
— Herbs or undershrubs. Flowers bisexual. Petals 4. Stamens 2, 4, or 6. Style 1, stigmas 1 or 2.—Ovary 2-locular. .... **Cruciferae**
646. Corolla imbricate, rarely valvate, then endosperm scanty or absent. Micropyle extrorse, rarely introrse. 647
— Corolla valvate. Endosperm copious. Micropyle introrse.—Leaves undivided. Stipules absent. Flowers actinomorphic, bisexual. Ovary 3- or 4-locular. Fruit a drupe. .... **Olacaceae**
647. Leaves not translucent-glandular-punctate, or rarely so, then stipules present.. 648
— Stipules absent. Leaves translucent-glandular-punctate..... **Rutaceae**
648. Calyx imbricate. ....:............... 649
— Calyx valvate.—Plants usually herbaceous. Leaves simple. Stipules present. Flowers in cymes. Stamens 5–10................ **Tiliaceae**

66

649. Stipules present, sometimes early fugacious. 650
— Stipules absent. 651
650. Sepals eglandular. Pedicels not articulate. Ovules anatropous. Embryo straight.—Medifixed hairs absent. .... **Zygophyllaceae**
— Sepals usually with large glands at base. Pedicels articulate. Ovules usually hemi-anatropous. Embryo usually curved.—Medifixed hairs present. .... **Malpighiaceae**
651. Leaves alternate. 652
— Leaves opposite or in whorls. (see 650). .... **Malpighiaceae**
652. Ovary 1 with 2 locules. 653
— Ovaries 3–5, free.—Flowers in umbels or in panicles. Petals 3–5. Stamens 6–10, appendiculate at base. Endosperm present or absent. .... **Simaroubaceae**
653. Flowers usually solitary. Petals 5. Stamens 5. Endosperm absent. Tropical America. (*Pelliceriaceae*). .... **Theaceae**
— Flowers in racemes. Petals 4, rarely 2. Stamens 4–6, rarely more. Endosperm ruminate. New Caledonia to New Guinea. (also included in *Aquifoliaceae*). .... **Sphenostemonaceae**
654. (641). Anthers with 2 longitudinal slits. 655
— Anthers with 1 apical pore.—Leaves undivided. Stipules usually absent. Flowers zygomorphic. .... **Polygalaceae**
655. Herbs, sometimes woody at base, or undershrubs. Stipules present. Fruit a 5-locular schizocarp, not winged, usually awned. Temperate parts.—Leaves pinnately partite to -compound, or digitately nerved.
                                                                    656
— Woody plants, rarely somewhat herbaceous, then leaves opposite or in whorls, simple and stipules absent. Fruit a capsule, or a berry, rarely a schizocarp, then 2- or 3-locular and often winged. (Sub-)tropics. 657
656. Flowers solitary, paired, or in umbels. Mericarps awned, very rarely not so, leaves then palmatinerved.—Lower cauline leaves opposite. (*Geranieae*). .... **Geraniaceae**
— Flowers in spikes or racemes. Mericarps unawned.—Lower cauline leaves alternate, pinnately partite to -compound. Greece to C. Asia. (*Biebersteiniaceae*). .... **Geraniaceae**
657. Leaves simple, undivided, usually opposite. Filaments connate at base only. 658
— Leaves pinnately compound, rarely 3-partite, usually alternate. Filaments connate into a tube for most of their length.—Stipules absent. (*Melioideae*). .... **Meliaceae**
658. Woody plants. Stipules usually present. Sepals imbricate, often with large glands. Petals imbricate.—Indument usually with medifixed hairs. .... **Malpighiaceae**

67

— Woody herbs or undershrubs. Stipules absent. Sepals valvate, eglandular. Petals contort.—Chile, S. Brazil. (*Vivianiaceae*).

Geraniaceae

659. (628). Ovules 2 per locule................................. 660
— Ovules 3–more per locule. 690
660. Stipules present, sometimes early fugacious. 661
— Stipules absent.. 677
661. Flowers unisexual.—Ovary 2-locular. Ovules erect. 662
— Flowers bisexual or polygamous.......................... 664
662. Petals 4, imbricate. Stamens alternipetalous.—Leaves undivided. 663
— Petals 4 or 5, valvate. Stamens epipetalous.—Leaves alternate. Fruit a berry. Endosperm copious.......................... **Vitaceae**
663. Leaves opposite. Fruit a berry. Endosperm absent. ... **Salvadoraceae**
— Leaves alternate. Fruit a drupe. Endosperm scanty..... **Celastraceae**
664. Calyx valvate. 665
— Calyx imbricate or apert. 668
665. Filaments free.—Ovules pendulous.......................... 666
— Filaments usually connate.—Endosperm present.............. 667
666. Stem usually herbaceous. Flowers in fascicles. Endosperm present.

Tiliaceae

— Stem woody. Flowers in panicles. Endosperm absent.

Dipterocarpaceae

667. Leaves alternate. Petals contort. Ovules ascending to patent.

Sterculiaceae

— Leaves opposite. Petals valvate. Ovules pendulous. (*Anopyxis*).

Rhizophoraceae

668. Stigma 1, undivided or lobed. 669
— Stigmas 2–5.—Flowers solitary, or in umbels, or in racemes, or in cymes. Fruit dehiscent, or a schizocarp, mericarps usually beaked.

Geraniaceae

669. Leaves compound. 670
— Leaves simple... 672
670. Inflorescences axillary or terminal. Stamens alternipetalous or more than the petals. 671
— Inflorescences usually leaf-opposed. Stamens epipetalous, 4 or 5.— Woody plants, usually climbing, then often with tendrils. Leaves usually digitately or 1-pinnately compound. Petals valvate. **Vitaceae**
671. Small, unarmed annuals. Leaves alternate. Sepals and petals 4. Stamens 6. Ovary 2-locular. Endosperm absent. Embryo curved. (*Oxystylidaceae*, also in *Cleomaceae*). .... **Capparaceae**
— Much-branched perennials or shrubs, often armed. Leaves opposite. Sepals and petals 5. Stamens 10. Ovary 5-locular. Endosperm present. Embryo straight. (*Fagonia, Plectrocarpa*). **Zygophyllaceae**

672. Anthers with longitudinal slits, rarely with pores, then ovary 2- or 3-locular. 673
— Anthers with apical pores. Ovary 4- or 5-locular.—Stamens 5, alternipetalous. . . . . **Ochnaceae**
673. Corolla imbricate or apert, rarely valvate, then stamens twice as many as the petals. 674
— Corolla valvate. Petals 4 or 5. Stamens epipetalous. . . . . **Vitaceae**
674. Leaves alternate. 675
— Leaves opposite, if alternate, stipules free and sepals 5.—Flowers solitary or in fascicles. Fruit a schizocarp. (*Fagonia, Viscainoa*).
Zygophyllaceae
675. Flowers in panicles or in fascicles. 676
— Flowers solitary.—Stipules intra-petiolary connate. Sepals 8–10, very unequal. New Caledonia. . . . . **Strasburgeriaceae**
676. Flowers in panicles. Fruit dry, indehiscent, rarely ultimately dehiscent, usually with an enlarged calyx.—Ovary 2- or 3-locular.
Dipterocarpaceae
— Flowers in fascicles. Fruit a septicidal capsule or a drupe.—Stem woody. Petals imbricate. Stamens as many as the petals, alternipetalous, or twice as many, obdiplostemonous. Anthers with longitudinal slits. (*Ixonanthaceae*). . . . . **Linaceae**
677. (660). Leaves not translucent-glandular-punctate, or rarely so, then either stamens connate at base, or less than the petals. 678
— Leaves translucent-glandular-punctate. Stamens as many as the petals or twice as many, free.—Stem woody. Anthers with longitudinal slits. . . . . **Rutaceae**
678. Stem woody, rarely only so at base, then stamens 5, or 8, or 10. 679
— Stem herbaceous, or woody at base only, then stamens 2, or 4, or 6.—Petals 4. Ovary 2-locular. . . . . **Cruciferae**
679. Stamens as many as the petals, alternipetalous, or more, or less. 680
— Stamens as many as the petals, epipetalous.—Calyx valvate. Petals 5, small, scale-like. Anthers usually with 2 apical pores. **Sterculiaceae**
680. Anthers with apical pores, or with poriform or transverse slits.— Filaments free. Anthers basifix. 686
— Anthers with longitudinal slits. 681
681. Leaves opposite. Filaments free. 682
— Leaves alternate, rarely opposite, then filaments more or less connate into a tube. 683
682. Petals contort. Stamens 8–10. Embryo curved. (*Vivianiaceae*).
Geraniaceae
— Petals not contort. Stamens 2(–4). Embryo straight. . . . . **Oleaceae**
683. Stamens 6–10, rarely less, then either leaves pinnately compound, or filaments more or less connate into a tube, or ovary 3–more-

69

locular, or sepals and petals 3 or 4. 684

— Leaves simple, entire. Sepals and petals 5. Stamens 5. Filaments at base connate into a ring. Ovary 2-locular. (*Ixonanthaceae: Cyrillopsis*). .... **Linaceae**

684. Sepals and petals usually 4 or 5, imbricate. 685

— Sepals and petals 3 or 4, valvate.—Twigs and petioles with a wavy, pale, sclerenchymatous ring around resinous ducts in transverse section. Filaments free............................... **Burseraceae**

685. Filaments usually connate into a tube.................. **Meliaceae**

— Filaments connate at base only.—Madagascar..(*Asteropeiaceae*). **Theaceae**

686. Ericoid (under-)shrubs. Flowers solitary, axillary.............. 687

— Shrubs or trees. Flowers in inflorescences.—Calyx and corolla imbricate. Anthers with 2 pores or slits. S.E. Asia................. 688

687. Leaves, young stems, and calyx with long, club-shaped glands. Calyx and corolla imbricate. Anthers with 2 apical pores or short slits.—Leaves alternate. S. Africa. .... **Roridulaceae**

— Plants without such glands. Calyx valvate, corolla induplicative-valvate. Anthers with 1 apical pore.—Leaves alternate, or opposite, or in whorls. Australia. .... **Tremandraceae**

688. Fertile stamens 2–5. Anthers with apical pores or transverse slits. 689

— Fertile stamens 10. Anthers with poriform, introrse slits.—Flowers in cymes. Ovary 3-locular. Ovules collateral. (*Sladeniaceae*). **Theaceae**

689. Flowers in panicles. Fertile stamens 2 or 3, epipetalous. Ovary 2- or 3-locular. Ovules serial. (*Meliosmaceae: Ophiocaryon*). .. **Sabiaceae**

— Flowers in racemes. Fertile stamens 5, alternipetalous. Ovary 5-locular. Ovules collateral...................... **Pentaphylacaceae**

690. (659). Placenta(s) central, or axillary, or apical. 693

— Placentas parietal.—Sepals and petals 4. Endosperm absent. Embryo curved. 691

691. Calyx 4- or 5-lobed. Petals 5. Ovary with a longitudinal false sept.—Herbs or shrubs. (*Astragalus*)...................... **Leguminosae**

— Sepals 4. Petals 4.—Endosperm absent. Embryo curved. 692

692. Filaments all equal in length.—Undershrubs........... **Capparaceae**

— Filaments unequal in length.—Corolla actinomorphic or radiate, not spurred. Stamens free or 4 pairwise connate; anthers 2-locular. Endosperm scanty or absent. Embryo large, curved......... **Cruciferae**

693. Stamens as many as the petals, or less. 694

— More stamens than petals. 704

694. Stamens as many as the petals, or less, then bracts of the sterile flowers (if any) not strongly modified. 695

70

— Petals 5. Stamens 3. Bracts of sterile flowers modified, saccate, pitcher-like, or spathulate, brightly coloured........ **Marcgraviaceae**

695. Anthers free.. 696

— Anthers connate around the ovary.—Herbs, rarely shrub-like. Flowers zygomorphic. Petals and stamens 5.............. **Balsaminaceae**

696. Flowers actinomorphic or nearly so......................... 697

— Flowers zygomorphic.—Woody plants. Petal and fertile stamen 1. C.-, S.-America................................... **Vochysiaceae**

697. Trees or shrubs, rarely herbs. Leaves simple, undivided, or digitately compound. Stipules usually present, sometimes absent. Flowers 3–5-merous.................................. 698

— Herbs. Leaves 3-foliolate. Stipules absent. Flowers 8-merous. C.-, S.-America..................................... **Tovariaceae**

698. Stamens epipetalous. Filaments usually connate............... 699

— Stamens alternipetalous. Filaments free.—Calyx imbricate. 700

699. Leaves simple. Anthers 2-locular, locules rarely sub-confluent at the apex.—Calyx valvate. .... **Sterculiaceae**

— Leaves digitately compound. Anthers 1-locular.—Stipules present.
Bombacaceae

700. Well-developed leaves present. Fruit a loculicid capsule or indehiscent.................................... 701

— Leaves reduced to minute scales. Fruit an apically irregularily septicid capsule.—Texas, Mexico. (*Canotia*, also in *Koeberliniaceae*).
Canotiaceae

701. Leaves either serrate or with 3 apical teeth. Ovary 3- or 5-locular.
702

— Leaves entire. Ovary 1- or 2-locular............... **Pittosporaceae**

702. Leaves either with 3 apical teeth, or glandular serrate, without long club-shaped glands. Ovary 3- or 5-locular. America or Australia. 703

— Leaves usually pinnatifid, with long club-shaped glands. Ovary 3-locular. S. Africa............................... **Roridulaceae**

703. Leaves with 3 apical teeth. Ovary 3-locular. S. America. (*Tribelaceae*). .... **Saxifragaceae**

— Leaves glandular-serrate. Ovary 5-locular. N.E. Australia. (*Abrophyllum*, *Cuttsia*, also included in *Escalloniaceae*) .... **Saxifragaceae**

704. (693). Stamens 6–10. 705

— Stamens 5.—Herbs or undershrubs. Flowers zygomorphic. Petals 3. Anthers connate. .... **Balsaminaceae**

705. Anthers with apical pores. 706

— Anthers with 2 longitudinal slits, sometimes poriform, but opening from the base up. 709

706. Anthers with 2 apical pores. Calyx and corolla imbricate........ 707

— Anthers with 1 apical pore. Calyx valvate, corolla induplicative-valvate.—Ovary 2-locular. Australia. . . . . **Tremandraceae**

707. Stipules absent. Anthers dorso-versatile. 708

— Stipules present. Anthers basifix-adnate.—Stem woody. Ovary 3–5-locular. . . . . **Ochnaceae**

708. Herbs. Ovary 5-locular. . . . . **Pyrolaceae**

— Woody plants. Ovary 3-locular. . . . . **Clethraceae**

709. Sepals valvate.—Leaves alternate, simple, rarely absent. Sepals free or connate. 710

— Sepals imbricate or apert.—Leaves compound, or simple, then if also alternate, either without stipules, or stamens free. 713

710. Filaments free. 711

— Filaments connate.—Stipules present. Sepals usually connate.

Sterculiaceae

711. Woody plants. Stamens 8. Ovary long-stipitate. Stigma usually sessile. Fruit a berry. Endosperm absent. Embryo curved. **Capparaceae**

— Herbs or undershrubs. Stamens 10. Ovary sessile, rarely with an androgynophore. Style present. Fruit a capsule. Endosperm present. Embryo straight.—Flowers yellow. 712

712. Staminodes absent. (*Corchorus*). . . . . **Tiliaceae**

— Staminodes present. (*Corchoropsis* also in *Sterculiaceae*). . . **Tiliaceae**

713. Stipules present.—Leaves not translucent-glandular-dotted. Stamens free. Anthers dorsifix. Endosperm present. 714

— Stipules absent. 715

714. Leaves compound. Flowers solitary, or in dichasia, or in fascicles. Aril absent.—Leaves usually opposite. Flowers 4- or 5-merous. Fruit a capsule or a berry. . . . . **Zygophyllaceae**

— Leaves simple. Flowers in spikes or racemes. Aril present.—Leaves alternate, serrate. Flowers 4-merous. Fruit a berry. S.E. Asia.

Stachyuraceae

715. Filaments free. 716

— Filaments more or less connate into a tube.—Anthers basifix.

Meliaceae

716. Leaves simple or rarely absent. 717

— Leaves compound, translucently-glandular-dotted. . . . . **Rutaceae**

717. Bracts of sterile flowers (if any) not strongly modified.—Shrubs. Leaves simple, undivided. 718

— Bracts of sterile flowers pitcher-like, saccate, or spurred, brightly coloured.—Anthers basifix, introrse. Ovary 2–many-locular. Tropical America. . . . . **Marcgraviaceae**

718. Stamens 6–10. Endosperm scanty or absent. 719

— Stamens 10. Endosperm copious.—Anthers dorso-versatile. Ovary 5-locular. W.-, C. China. (*Clematoclethra*). . . . . **Actinidiaceae**

719. Anthers basifix. Ovary sessile. (incl. *Asteropeiaceae*). . . . . . **Theaceae**

— Anthers dorsifix. Ovary short- to long-stipitate. (incl. *Koeberliniaceae*: *Koeberlinia*). . . . . **Capparaceae**

720. (551). Ovary 1, undivided or lobed. 721
— Ovaries 2 – more, free, or connate at base. 770

721. Ovary 1-locular, sometimes incompletely more-locular. 722
— Ovary completely 2 – more-locular, or nearly so. 740

722. Ovules 3 – more, or flowers unisexual. 723
— Ovules 1 or 2. 725

723. Petals of all flowers free. 724
— Petals of the male flowers connate.—Trees. Leaves digitately lobed to -compound. Flowers unisexual or polygamous. . . . . **Caricaceae**

724. Petals 4 – 7. 730
— Petals 2.—Stem herbaceous, sometimes woody at base. Leaves alternate, sometimes in tufts. Stigmas 4, free. Ovules parietal. Endosperm absent. Embryo curved. **Resedaceae**

725. Ovules erect or ascending.—Flowers bisexual, rarely unisexual. 726
— Ovules pendulous or descending.—Leaves alternate. Filaments more or less connate. Styles 3 or 4. Fruit a drupe. 729

726. Leaves alternate. Stipules absent. Ovule 1.—Stamens 5. 727
— Leaves opposite. Stipules scarious, or an inter-petiolary ridge present. Ovules 2.—Herbs or undershrubs. Calyx imbricate. Fruit a capsule, or dry, indehiscent. . . . . **Caryophyllaceae**

727. Shrubs, rarely herbs or trees. Stamens either alternipetalous, or more than the petals. Styles 3, or 1 with 3 stigmas. 728
— Herbs, undershrubs, or climbers. Stamens epipetalous. Styles 5.— Calyx plicate. Aril absent. (*Plumbagineae*). . . . . **Plumbaginaceae**

728. Stipules absent. Petals, stamens 5.—Tropical Africa. (*Stapfiella*, also in *Flacourtiaceae*). . . . . **Turneraceae**
— Stipules present, usually connate into a sheath. Petals 3, sometimes 4. Stamens or staminodes 3 – 9. (*Coccolobeae*). . . . . **Polygonaceae**

729. Stipules absent. Flowers unisexual. Stamens 1 – 5. Anthers adnate. Embryo curved. . . . . **Menispermaceae**
— Stipules present. Flowers bisexual. Stamens 10. Anthers versatile. Embryo straight. . . . . **Erythroxylaceae**

730. Staminodes, if present, shorter than the petals, or not both filiform and pubescent. 731
— Staminodes filiform, 5, longer than the petals, densely pubescent.— Trees. Ovary 1-locular with a slender central column and 3 or 4 carpels. Ovules pendulous from the apex of the locule, close to the column, 6 – 8. W. Africa. . . . . **Medusandraceae**

731. Placentas several, parietal, rarely basal, then all leaves radical. Embryo straight. 732
— Leaves opposite, undivided. Placenta 1, basal or central. Embryo

more or less curved.—Herbs or undershrubs. Flowers bisexual. Sepals 4 or 5, imbricate. . . . . **Caryophyllaceae**

732. Flowers unisexual. 733
— Flowers bisexual or polygamous.—Herbs or (under-)shrubs. 735

733. Shrubs or undershrubs, climbing with hooks or tendrils. 734
— Erect shrubs or trees without tendrils or hooks.—Leaves undivided. Stipules present. . . . . **Flacourtiaceae**

734. Stipules present, usually small and fugacious. Midribs of leaves not excurrent into hooks. . . . . **Passifloraceae**
— Stipules absent. Midrib of at least some leaves excurrent into 2 re-curved hooks.—Seeds large, discoidal. . . . . **Dioncophyllaceae**

735. Sepals 4 – 7. 736
— Sepals 2 or 3.—Leaves alternate, usually incised. Stipules absent. Sepals free, imbricate. Petals 4 – 6. Style 2- or 3-partite. **Papaveraceae**

736. Sepals free.—Leaves alternate. Stipules absent. 737
— Sepals connate at base. 738

737. Leaves often scale-like. Placentas usually basal. Seeds hairy.
Tamaricaceae
— Leaves entire, crenate, or lobed, or pinnatifid. Placentas parietal. Seeds arillate. . . . . **Turneraceae**

738. Leaves alternate, or all radical, rarely in whorls. Calyx imbricate.
739
— Leaves opposite. Calyx valvate. . . . . **Frankeniaceae**

739. Plants insectivorous. Petals imbricate. Anthers extrorse, 4 – 10. Seeds not arillate. . . . . **Droseraceae**
— Plants not insectivorous. Petals contort. Anthers introrse, 5. Seeds arillate.—Herbs or undershrubs. . . . . **Turneraceae**

740. (721). Ovules 1 or 2 per locule. 741
— Ovules 3 – more per locule. 760

741. Ovules patent or ascending. 742
— Ovules pendulous or descending. 750

742. Leaves opposite.—Flowers unisexual or polygamous, 4- or 5-merous. Anthers 2-locular. 743
— Leaves alternate. 744

743. Filaments free. Styles 2. . . . . **Aceraceae**
— Filaments connate. Styles 4 or 5. . . . . **Guttiferae**

744. Anthers 1- or 3 – more-locular. 745
— Anthers 2-locular. 746

745. Sepals 5. Anthers 1-locular. . . . . **Malvaceae**
— Sepals 3. Anthers 3 – more-locular. . . . . **Bombacaceae**

746. Flowers bisexual or polygamous. Sepals 5, connate at base. Stamens 5 – 10. 747
— Flowers unisexual. Sepals 2 or 3, free. Stamens 2 or 3.—Ericoid

74

undershrubs. Stipules absent. Sepals imbricate. Petals 2 or 3. Filaments free. Ovule 1 per locule. . . . . **Empetraceae**

747. Stipules absent. Petals imbricate or valvate. 748
— Stipules present. Petals contort.—Filaments connate. Style 1. Ovules 2 per locule. . . . . **Sterculiaceae**

748. Leaves simple. Flowers bisexual. Filaments connate at base. Styles 2–7, free. Endosperm mealy. 749
— Leaves pinnate. Flowers polygamous or unisexual. Filaments free. Style 1. Stigma 5-lobed. Endosperm absent or scanty.—Sepals connate at base. Ovule campylotropous. Madagascar. (*Ptaeroxylaceae*: *Cedrelopsis*). . . . . **Meliaceae**

749. Sepals free. Ovule erect, epitropous.—Leaves fleshy. Stamens 10. Aril present. C. America. (*Stegnospermataceae*). . . . . **Phytolaccaceae**
— Sepals connate at base. Ovule erect or patent, apotropous.—Aril present, Australia (*Macarthuria*), or aril absent, Africa to India (*Limeum*). . . . . **Aizoaceae**

750. Flowers unisexual. 751
— Flowers bisexual or polygamous. 752

751. Petals imbricate. Ovules epitropous, usually 2 per locule, collateral, usually with a caruncle.—Stipules usually present. . . . **Euphorbiaceae**
— Petals valvate. Ovule 1 per locule, more or less campylotropous, without a caruncle.—Stipules absent. New Caledonia. (*Phelline*, also in *Aquifoliaceae*). . . . . **Phellinaceae**

752. Leaves usually alternate. Calyx rarely with glands outside, sometimes with apical calli (*Anisadenia*). Ovules anatropous, rarely atropous or campylotropous. Endosperm present, sometimes scanty. 753
— Leaves usually opposite. Calyx usually with glands outside. Ovule 1 per locule, more or less hemitropous. Endosperm absent.—Woody plants, rarely undershrubs. Petals usually clawed, dentate, or fimbriate. Leaves undivided. Ovary usually lobed. Styles 2–4.
Malpighiaceae

753. Filaments free. 754
— Filaments connate at base. 756

754. Leaves compound, imparipinnate.—Flowers 5-merous. Calyx slightly connate, slightly imbricate. Madagascar. (*Ptaeroxylaceae*: *Cedrelopsis*). . . . . **Meliaceae**
— Leaves simple. 755

755. Sepals imbricate, or calyx 4- or 5-dentate, or nearly entire. Stamens 4 or 5. Stigma sessile, capitate, or discoid.—Flowers solitary, or in fascicles. Endosperm present. **Aquifoliaceae**
— Sepals valvate. Stamens 10, or 5, then style 2- or 3-lobed, or long and undivided.—Leaves usually alternate. . . . . **Cyrillaceae**

756. Leaves simple, undivided. Ovary undivided. 757

— Leaves usually compound, sometimes unifoliolate. Ovary lobed.—
Styles 5, free. (incl. *Averrhoaceae*). .... **Oxalidaceae**
757. Petals inside without appendages........................... 758
— Petals inside with inflated or scale-like appendages.—Woody plants.
Stipules present. Flowers solitary or in fascicles. Stamens 10. Styles
3 or 4. Fruit a drupe......................... **Erythroxylaceae**
758. Leaves alternate. Flowers without an epicalyx. 759
— Leaves opposite. Flowers with an epicalyx.—Flowers solitary or in
fascicles. Stigmas usually 3. Argentine, Chile. (*Ledocarpaceae*:
Wendtia). .... **Geraniaceae**
759. Petals longer than the sepals, usually contort and clawed. Staminal
tube usually with alternipetalous glands, when absent plants either
herbaceous, or woody and climbing with hooks. Styles and/or style-
branches filiform. Endosperm present. Embryo straight.... **Linaceae**
— Petals ca. as long as the sepals, imbricate, not clawed. Staminal
tube eglandular. Styles 3, very short. Endosperm scanty. Embryo
curved.—Woody plants, not climbing with hooks. Madagascar.
(*Asteropeiaceae*). .... **Theaceae**
760. (740). Styles or sessile stigmas free......................... 761
— Styles connate at base. 762
761. Flowers bisexual. 763
— Flowers unisexual or polygamous.—Woody plants. Leaves opposite,
with translucent-glandular dots or lines. Stipules absent. Filaments
connate. Endosperm absent. .... **Guttiferae**
762. Flowers bisexual. Petals always free. 768
— Flowers polygamous or unisexual. Male flowers with connate petals.—Trees. Leaves digitately lobed or divided, terminally tufted.
Anthers with 2 longitudinal, introrse slits. .... **Caricaceae**
763. Filaments connate at base.—Leaves alternate. Endosperm present.
764
— Filaments free. 765
764. Leaves usually compound, sometimes unifoliolate. Ovules axillary.
(incl. *Averrhoaceae*). .... **Oxalidaceae**
— Leaves simple. Ovules basal.—Australia, New Caledonia. (*Macarthuria*). .... **Aizoaceae**
765. Stipules absent. Endosperm scanty to copious. 766
— Stipules present. Endosperm absent or nearly so.—Herbs or undershrubs. Leaves opposite or in whorls, simple or undivided.
Elatinaceae
766. Herbs. Carpels connate up to the middle...................... 767
— Undershrubs. Carpels connate up to the stigmas.—Flowers solitary.
Sepals, petals 5. Endosperm scanty. S. America. (*Balbisia*, also in
Ledocarpaceae). .... **Geraniaceae**

767. Leaves opposite, undivided. Calyx 4–6-partite. Petals 4–6. Endosperm scanty.—Flowers in cymes.....................**Crassulaceae**
— Leaves alternate, partly lobed. Flowers 5-merous. Endosperm copious.—Tepals 5, corolloid. Nectaries lobed. (*Helleboraceae*).
Ranunculaceae
768. Calyx imbricate.—Leaves undivided. Stipules absent. 769
— Calyx valvate.—Calyx 5-lobed or -partite. Filaments more or less connate. .... **Sterculiaceae**
769. Calyx 5-partite. Filaments free or nearly so. Anthers with 2 apical pores.—Stamens 10................................**Clethraceae**
— Sepals 5, free. Filaments distinctly connate at base. Anthers with slits.—Stamens 8–10. Madagascar. (*Asteropeiaceae*). ..... **Theaceae**
770. (720). Ovule 1 per carpel.................................771
— Ovules 2–more per carpel. 786
771. Ovule ascending, basal................................772
— Ovule descending. 778
772. Herbs...773
— Woody plants.—Anthers introrse or latrorse. Endosperm copious.
775
773. Leaves alternate or radical. Carpels indehiscent...............774
— Leaves usually opposite. Carpels dehiscent.—Anthers dorsifix, introrse. Carpels 3–9. Endosperm scanty..............**Crassulaceae**
774. Anthers introrse. Carpels 3–5. Endosperm absent.—Leaves incised or compound. Sepals and petals 3–5. Anthers versatile. ... **Limnanthaceae**
— Anthers extrorse or latrorse. Carpels numerous. Endosperm copious. (*Ranunculoideae*). .... **Ranunculaceae**
775. Petals up to 6...776
— Petals many.—Anthers introrse or latrorse. 777
776. Ovaries in a whorl, 5–20. Fruit consisting of ventrally dehiscing follicles. (also in *Magnoliaceae*). .... **Illiciaceae**
— Ovaries in a spiral, many. Fruit either indehiscent, or consisting of dorsally dehiscing follicles. .... **Magnoliaceae**
777. Petals 6. Anthers extrorse. Carpels rarely dehiscent.—Sepals 3. Anthers adnate. .... **Annonaceae**
— Petals 3–5. Anthers introrse or latrorse. Carpels dehiscent. (*Didesmandra, Hibbertia*). .... **Dilleniaceae**
778. Flowers unisexual.—Leaves alternate.......................779
— Flowers bisexual or polygamous............................781
779. Leaves simple..780
— Leaves pinnately compound.—Tree. Stipules absent. Flowers in panicles. Stamens epipetalous, as many as the petals, free, 4(–7). Fruit a drupe. Peru, Brazil. (*Picrolemma*)..........**Simaroubaceae**
780. Shrubs, rarely herbs or undershrubs. Stipules absent. Flowers not in

globose capitules. Stamens epipetalous, rarely less or more than the petals. Mericarps drupaceous. .... **Menispermaceae**
— Trees. Stipules present. Flowers in globose capitules. Stamens alternipetalous, nearly free, 3–8. Carpels 3–8, nut-like. .... **Platanaceae**
781. Leaves alternate or radical.—Stipules absent. Ovaries free. Endosperm present................................... 782
— Leaves opposite or in whorls. 784
782. Climbers or herbs. Leaves not terminally tufted. 783
— Trees. Leaves terminally tufted.—Stamens, staminodes, ovaries 5. Indomalesia. (*Eurycoma*). .... **Simaroubaceae**
783. Woody climbers. Anthers introrse. Ovaries 3–12.—Sepals and petals 6. Stamens 6–8. (*Parabaena, Tiliacora*)..... **Menispermaceae**
— Herbs. Anthers extrorse or latrorse. Ovaries many. (*Ranunculoideae*). .... **Ranunculaceae**
784. Stipules absent. Filaments free.—Anthers extrorse............ 785
— Stipules present. Filaments connate.—Woody plants. Petals 5. Carpels 3. Endosperm absent. .... **Malpighiaceae**
785. Shrubs. Petals 5, greenish. Stamens 10. Mericarps nut-like.— Anthers introrse. Carpels 5–10. .... **Coriariaceae**
— Herbs. Petals 3–9, coloured. Stamens 3–9. Carpels follicular.— Carpels 3–9. Endosperm scanty. .... **Crassulaceae**
786. (770). Ovules 2 per carpel................................ 787
— Ovules 3–more per carpel 796
787. Herbs.—Flowers 3–5-merous............................. 788
— Trees or shrubs, rarely undershrubs.—Leaves alternate, rarely in whorls. 789
788. Terrestrial plants. Leaves opposite. .... **Crassulaceae**
— Aquatics. Leaves alternate, the floating leaves peltate, submerged leaves dissected. (*Cabomba: Cabombaceae*). .... **Nymphaeaceae**
789. Ovules descending..................................... 790
— Ovules ascending..................................... 793
790. Sepals connate..................................... 791
— Sepals free.—Endosperm present. 792
791. Leaves translucent-glandular-punctate, usually compound. Stamens 3–5, as many as the petals. Carpels 2–5. Endosperm present. Embryo straight..................................... **Rutaceae**
— Leaves not translucent-glandular-punctate, undivided. Stamens 10, at least twice the number of the petals. Carpels 5. Endosperm absent. Embryo curved.—Flowers bisexual. Fruits drupaceous. (also in *Simaroubaceae*). .... **Surianaceae**
792. Leaves pinnately compound. Flowers in panicles, polygamous. Filaments free. Carpels 5–15. Fruit a dry follicle. (*Helleboraceae: Xanthorrhiza*)................................. **Ranunculaceae**

— Leaves simple. Flowers solitary, unisexual. Filaments connate or coherent at base. Carpels many. Fruit drupe-like. (*Schisandraceae*).

Magnoliaceae

793. Leaves simple, undivided, sometimes absent. Ovules anatropous. 794

— Leaves compound. Ovules atropous, collateral......... **Connaraceae**

794. Anthers adnate. Endosperm copious. 795

— Anthers versatile. Endosperm absent. (also in *Simaroubaceae*).

Surianaceae

795. Sepals 3. Petals 6. Anthers extrorse. .... **Annonaceae**

— Sepals 5. Petals 3–5. Anthers extrorse or latrorse. .... **Dilleniaceae**

796. (786). Anthers adnate or basifix, extrorse, rarely introrse or latrorse. 787

— Anthers dorsoversatile or basiversatile, introrse................. 802

797. Leaves compound.—Flowers unisexual or polygamous. Stamens 6.

Lardizabalaceae

— Leaves simple.. 798

798. Style(s) present. 799

— Stigma subsessile... 801

799. Petals 6.—Endosperm ruminate..................... **Annonaceae**

— Petals 3–5. 800

800. Ovules 3–15 per locule. Aril present. (*Hibbertia*)...... **Dilleniaceae**

— Ovules many per locule. Aril absent.—Tasmania. (*Tetracarpaeaceae*).. **Saxifragaceae**

801. Wood with vessels. Twigs on cross-section with a regular pattern of radial medullary rays, dilating in the bark. Leaves hairy or glabrous. Calyx either with distinct lobes or sepals free. Endosperm ruminate.

Annonaceae

— Wood without vessels. Twigs without such medullary rays. Leaves glabrous. Calyx either calyptrate and caducous, or persistent, then cup- or saucer-shaped, entire or ruptured into more or less irregular 'lobes'. Endosperm not ruminate. (*Drimys, Pseudowintera*).

Winteraceae

802. Stipules absent.—Anthers with 2 longitudinal slits. Carpels dehiscent. 803

— Stipules present. 804

803. Filaments free. Endosperm scanty or absent.......... **Crassulaceae**

— Filaments connate at base. Endosperm copious.—Calyx 5-partite. Brazil. (*Eichleria*)............................... **Oxalidaceae**

804. Calyx spatha-like. Anthers with 1 longitudinal slit. Staminodes corolloid. .... **Sterculiaceae**

— Sepals 5, free. Anthers with 2 longitudinal slits. Staminodes absent. .... **Simaroubaceae**

805. (550). Ovary 1, undivided, or lobed. 806

— Ovaries 2–more, free, or connate at base, or connate by the styles
only. 912
806. Ovary 1-locular, sometimes incompletely more-locular. 807
— Ovary 2–more locular or nearly so. 845
807. Ovules basal or nearly so. 808
— Ovules parietal or central. 813
808. Styles 2–5. Stigmas 2–5. 809
— Style 1. Stigmas 1–4. 811
809. Non-insectivorous plants, usually woody. 810
— Insectivorous herbs.—Leaves with glandular hairs or marginal
bristles. Flowers in cincinnate cymes. Styles 5, free, or connate
almost up to the 5 free stigmas. Placentas not extending to the apex
of the locule. . . . . **Droseraceae**
810. Leaves scale-like. Stipules absent. Flowers solitary. Placentas ex-
tending almost to the apex of the locule. (*Reaumurieae*).
Tamaricaceae
— Leaves well-developed. Stipules present, usually connate into a
sheath. Flowers in long racemes or spikes. Ovule 1, basally attached
with a long funicle.—Tropical America. (*Symmeria*). . **Polygonaceae**
811. Woody plants, rarely herbaceous, then leaves herbaceous or
coriaceous. Embryo straight. 812
— Succulent herbs or shrubs with succulent leaves. Embryo curved.—
Sepals 2(–8). Petals 3–15. Style-branches 2–8. Endosperm present,
usually thin. . . . . **Portulacaceae**
812. Leaves alternate, rarely opposite, then, as usual, fruit follicular and
seeds arillate. Endosperm copious. **Dilleniaceae**
— Leaves opposite. Fruit a drupe or a berry. Endosperm absent.—
Leaves with translucent-glandular lines or dots. . . . . **Guttiferae**
813. Placenta parietal. 814
— Placenta central, free.—Corolla valvate. Ovules 3, pendulous. Fruit
a drupe. . . . . **Olacaceae**
814. Placenta 1. 815
— Placentas 2–more. 825
815. Leaves simple, rarely incised or compound, then stipules absent. 816
— Leaves compound, or reduced to a broadened, leaf-like petiole.—
Stipules more or less distinct. Ovules 2–more. . . . . **Leguminosae**
816. Calyx valvate. Corolla valvate or imbricate. 817
— Calyx and corolla imbricate or apert. 818
817. Stipules present. Calyx 5-fid. Petals 5, imbricate. Anthers 1-locular.
Ovule 1, pendulous.—Filaments connate. . . . . **Malvaceae**
— Stipules absent. Sepals 3, valvate. Petals 6, usually valvate, some-
times imbricate. Anthers adnate, 2-locular. Ovules 2 or more.
Annonaceae

sepals 2, then either stem woody, or sepals connate at base. 833
— Sepals 2, free, rarely 3, then stem, as usual, herbaceous and petals
6.—Leaves alternate............................... **Papaveraceae**
833. Sepals usually 4, rarely 3, then either stem woody or petals 3–5,
rarely sepals 2, then either stem woody, or sepals connate at base,
rarely sepals 5. 834
— Sepals 3. Stem woody. Leaves alternate. Petals 12 or 13, imbricate.
Stamens 12. Staminodes 11 or 12. Ovary open along ventral suture
in very young stages.—Fiji Islands.................. **Degeneriaceae**
834. Leaves opposite or in whorls. 835
— Leaves alternate. 838
835. Leaves without translucent-glandular dots or lines. Endosperm
mealy... 836
— Leaves with translucent-glandular dots or lines. Endosperm absent.—Stipules absent. Flowers actinomorphic. Stigmas usually 2–5.
Embryo usually straight............................... **Guttiferae**
836. Woody plants. Sepals 4, valvate, free. Petals 4, imbricate. Seeds
arillate, stellately hairy. Endosperm scanty. S. Africa. (*Pseudosco-
lopia*).. **Flacourtiaceae**
— Herbs or small shrubs. Sepals 3 or 5–7. Petals 5–7, when 4 imbricate. Seeds exarillate, glabrous. Endosperm mealy............ 837
837. Sepals 3 or 5, contort, free. Petals 5, contort. .... **Cistaceae**
— Sepals 6 or 7, induplicative-valvate, connate into a tube. Petals 4–7,
imbricate..................................... **Frankeniaceae**
838. Ovary sessile or subsessile................................ 839
— Ovary usually long-stipitate.—Stigma 1, usually sessile. Endosperm
absent. Embryo curved. (incl. *Cleomaceae*: *Tetratelia*). **Capparaceae**
839. Sepals valvate.—Indument usually stellate................... 840
— Sepals imbricate, or contort, or apert...................... 842
840. Inflorescences terminal or axillary......................... 841
— Racemes opposite to the leaves.—Leaves crenate. Petals 5, without
scales at base. Ovary slightly stipitate. Australia (?, once found).
(*Nettoa*).. **Tiliaceae**
841. Leaves entire to serrate. Bracteoles present, minute. Petals 3–5,
without a scale at base. Fruit subsessile, a berry or a capsule.
Embryo straight.—Placentas 2–8. S. America. (*Banara*, *Pineda*,
also in *Flacourtiaceae*). .... **Tiliaceae**
— Leaves sinuately lobed. Petals 4, with a hairy scale at base. Fruit
stipitate, swollen with constrictions. Embryo curved.—Placentas 2.
New Caledonia. (*Oceanopapaver*, also in *Capparaceae*).... **Tiliaceae**
842. Petals contort.. 843
— Petals imbricate or valvate.—Woody plants. Endosperm fleshy.
Embryo straight................................ **Flacourtiaceae**

82

843. Plants erect. Leaves without climbing-hooks. Sepals free, or connate at base only. Seeds ripening within the developing fruit. Embryo curved. 844
— Soft-wooded lianas. Midrib of leaves excurrent into 2 recurved hooks. Sepals connate into a 5-dentate tube. Fruit a very early dehiscent capsule, the ovules ripening on elongated, rigid funicles into large discoidal seeds. Embryo straight. Tropical W. Africa. (*Dioncophyllum*). . . . . **Dioncophyllaceae**
844. Herbs or smallish shrubs. Sepals contort. Ovary strictly 1-locular. Placentas 3 – 10, each with 2 – many, usually atropous ovules. Endosperm mealy. . . . . **Cistaceae**
— Large shrubs or trees. Sepals imbricate. Ovary 3-locular at base. Placentas 3, each with 2 anatropous ovules. Endosperm scanty to absent.—Tropical Africa. (*Marquesia*). . . . . **Dipterocarpaceae**
845. (806). Ovule 1 per locule. 846
— Ovules 2 – more per locule. 861
846. Flowers unisexual. Ovary 2 – 4-locular. Ovule pendulous. Endosperm present. 847
— Flowers bisexual or polygamous, rarely unisexual, then either ovary 5 – 10-locular, or ovule ascending. 848
847. Trees or shrubs. Male flowers with petals. Female flowers with staminodes. Sepals 4, valvate. Petals 4. Stamens 15 – more. Ovary 4-locular. Ovule without a caruncle. Peru, Brazil. (*Hydrogaster, Vasivaea*). . . . . **Tiliaceae**
— Plant otherwise. Ovule usually with a caruncle. . . . . **Euphorbiaceae**
848. Calyx valvate.—Leaves alternate. 849
— Calyx imbricate or apert, rarely closed or dome-shaped. 853
849. Stipules present, sometimes early fugacious. 850
— Stipules absent.—Carpels many, more or less connate. . **Annonaceae**
850. Filaments connate into several bundles or free. 868
— Filaments connate into 1 bundle. 851
851. Anthers with 2 slits. 852
— Anthers with 1 slit. . . . . **Malvaceae**
852. Calyx 3-lobed with an epicalyx. Ovary 2- or 3-locular. Stylebranches 2 or 3. . . . . **Bombacaceae**
— Calyx 5-partite, epicalyx absent. Ovary 5 – 10-locular. Styles 5 – 10.
Sterculiaceae
853. Trees or shrubs. 854
— Herbs or undershrubs.—Leaves alternate, divided. Stipules absent. Filaments free. . . . . **Cruciferae**
854. Leaves compound. 855
— Leaves simple, undivided. 856
855. Leaves digitately compound.—Styles 4 – more, free. Endosperm absent.

Ovules ascending. Tropical America. (*Caryocar*)....... **Caryocaraceae**

— Leaves pinnately compound.—Leaves alternate. Stipules absent. Filaments connate into a tube......................... **Meliaceae**

856. Leaves alternate. 857

— Leaves opposite.—Endosperm absent. 860

857. Stipules absent.. 858

— Stipules present.. 859

858. Corolla valvate.—Filaments free or nearly so. Ovules pendulous.

Olacaceae

— Corolla imbricate. (*Ternstroemiaceae*). .... **Theaceae**

859. Anthers adnate. Ovules ascending. Endosperm absent... **Ochnaceae**

— Anthers versatile. Ovules pendulous. Endosperm usually present. (*Nitraria*)..................................... **Zygophyllaceae**

860. Stipules absent. Ovules ascending. .... **Guttiferae**

— Stipules present. Ovules pendulous.—Styles 3. .... **Malpighiaceae**

861. (845). Calyx valvate.—Stipules present. 862

— Calyx imbricate or apert, rarely closed....................... 872

862. Ovary sessile or nearly so, when stipitate petals 5. Ovules usually axillary... 863

— Ovary usually long-stipitate. Petals 4. Ovules inserted on the sept.— Stigma, usually sessile. Endosperm absent or nearly so. Embryo curved. .... **Capparaceae**

863. Filaments free, or connate into several bundles. 864

— Filaments all connate into 1 bundle........................... 869

864. Flowers not lepidote outside, epicalyx absent. Anthers 2-locular, locules sometimes confluent at the apex.................... 865

— Flowers lepidote outside. Epicalyx 2–5-lobed. Anthers 1–more-locular, apically dehiscent. .... **Bombacaceae**

865. Petals calycoid or incised, usually sessile with a broad base, pubescent outside, valvate or induplicative-valvate, rarely imbricate, never contort. Filaments free. 866

— Petals corolloid, margin entire, rarely incised, then filaments connate into several bundles; base attenuate, glabrous, imbricate, usually contort, rarely valvate, then filaments connate into several bundles. 867

866. Anthers narrow, apically dehiscent.—Trees or shrubs. Ovules descending, or 1 descending and 1 ascending. .... **Elaeocarpaceae**

— Anthers broad, longitudinally dehiscent............. **Flacourtiaceae**

867. Ovary 3-locular. Ovules 2 per locule, descending.—Resinous trees. Flowers in panicles. .... **Dipterocarpaceae**

— Ovary 2-, or 4–more-locular, rarely 3-locular then ovules either many or ascending, rarely with 2 descending ovules, then herbs or undershrubs. 868

868. Staminodes present. (incl. *Nesogordonia*, also placed in *Tiliaceae*).

Sterculiaceae

— Staminodes absent.................................. **Tiliaceae**

869. Corolla contort.—Petals 5................................. 870

— Corolla valvate.—Anthers with 4 pores............. **Flacourtiaceae**

870. Anthers 1-locular, with 1 slit, rarely 2–more-locular, then epicalyx present and flowers with stiff scales......................... 871

— Anthers 2-locular, with 2 slits or pores. Epicalyx absent, rarely present (*Dombeya*), then leaves simple and pollen spiny.

Sterculiaceae

871. Pollen spiny.—Leaves simple. Anthers 1-locular. .... **Malvaceae**

— Pollen smooth, rarely reticulate or pusticulate.—Trees. **Bombacaceae**

872. (861). Stipules present, sometimes early fugacious. 873

— Stipules absent or very minute............................. 889

873. Leaves opposite.. 874

— Leaves alternate or all radical. 876

874. Style undivided. 875

— Styles 2–more, free.—Shrubs or trees...................... 888

875. Filaments connate at base. Endosperm absent.—Africa to India. (*Monsonia, Sarcocaulon*)........................... **Geraniaceae**

— Filaments free. Endosperm present..................... **Cistaceae**

876. Styles 3–more, free. 877

— Style 1, stigmas 1–several............................... 878

877. Ovary lobed. Ovules many per locule................ **Dilleniaceae**

— Ovary undivided. Ovules 2 per locule. S. America. (*Roucheria*).

Linaceae

878. Ovary sessile or nearly so, rarely stipitate, then anthers adnate and embryo straight. Ovules axillary............................. 879

— Ovary usually long-stipitate. Anthers dorso-versatile. Ovules usually inserted on the sept. Embryo curved.—Stigma 1, usually sessile.

Capparaceae

879. Ovules 2 per locule, ascending, or more, then sometimes descending.. 880

— Ovules 2 per locule, descending or patent..................... 887

880. Calyx apert, or closed, or valvate, rarely slightly imbricate...... 881

— Calyx distinctly imbricate. 883

881. Filaments connate. Anthers usually with 1 slit. 882

— Filaments free. Anthers with 2 longitudinal slits.—Herbs. Leaves irregularily multifid. Sepals nearly free to base. (*Peganum*).

Zygophyllaceae

882. Leaves usually digitately compound or lobed. Pollen smooth, rarely reticulate or pusticulate.—Trees.................... **Bombacaceae**

— Leaves simple, pinnately or digitately nerved. Pollen spiny.—Calyx

with a nearly entire margin. . . . . **Malvaceae**
883. Anthers with apical pores. 884
— Anthers with longitudinal slits.—Leaves undivided. 885
884. Leaves undivided or pinnately compound. Filaments short. Embryo
straight. . . . . **Ochnaceae**
— Leaves lobed or digitately compound. Filaments long. Embryo
curved. . . . . **Cochlospermaceae**
885. Filaments connate at base. Ovary completely locular. 886
— Filaments free. Ovary incompletely locular.—Leaves without translucent-glandular lines or dots. Ovules ascending. Fruit septicide.
Embryo curved. . . ., ... **Cistaceae**
886. Ovules descending, many. Embryo large, straight.—Fruit septicide.
(incl. *Mahurea*, also in *Bonnetiaceae* or *Theaceae*). . . . . **Guttiferae**
— Ovules erect, basal, 3 or 7 – 9 per locule. Embryo minute.—Fruit indehiscent, globose or kidney-shaped, densely muricate. Madagascar.
(*Sphaerosepalaceae*). . . . . **Cochlospermaceae**
887. Trees or shrubs. Flowers in spikes, or in racemes, or in panicles.
Ovary usually 3-locular, rarely 4- or 5-locular (*Pakaraimaea*).—
Anthers basifix-adnate and plants from S.E. Asia (*Dipterocar-
poideae*), or more or less basiversatile and plants from Africa (*Marquesia, Monotes*), or S. America (*Pakaraimaea*). . . **Dipterocarpaceae**
— Herbs. Flowers solitary or in umbels. Ovary 5-locular.—Anthers
versatile. (*Monsonia, Sarcocaulon*). . . . . **Geraniaceae**
888. (874). Flowers large, solitary. Stamens very many. Ovules several
per locule. Endosperm present. S. temperate. . . . . **Eucryphiaceae**
— Flowers small, in racemes or in panicles. Stamens 15 – 30. Ovules 2
per locule. Endosperm absent. Tropical S. America. . . . . **Quiinaceae**
889. (872). Leaves not tubular. 890
— Leaves tubular.—Insectivorous herbs. Leaves radical. Flowers 5-
merous. Endosperm copious. America. . . . . **Sarraceniaceae**
890. Sepals 4 – more, rarely 2 or 3, then either plant woody, or petals 3
or 5. 891
— Sepals 2 or 3. Petals 4 or 6.—Herbs. Flowers solitary. Endosperm
copious. . . . . **Papaveraceae**
891. Leaves compound, rarely lobed, then sepals 5, free and petals 5 or 8
modified into nectaries with lids. 892
— Leaves simple. 896
892. Leaves digitately compound. 893
— Leaves pinnately compound or lobed. 894
893. Ovary distinctly stipitate, 2 – 6-locular. . . . . **Capparaceae**
— Ovary sessile, 8 – 20-locular.—Leaves translucent-glandular-punc-
tate. . . . . **Rutaceae**

894. Herbs. Flowers solitary or in cymes. Sepals free. Ovules many per locule. 895
— Woody plants. Flowers in spikes, or in racemes, or in panicles. Sepals connate. Ovules 2 per locule.—Leaves cauline. . . . . **Meliaceae**
895. Leaves radical. Flowers in cymes. Petals 5, contort, not modified into nectaries. Embryo coiled. Andes. (*Hypseocharitaceae*).

Oxalidaceae
— Leaves cauline. Flowers solitary. Petals 4 or 8, imbricate, modified into nectaries with lids. Embryo straight. Eurasia. (*Helleboraceae-Nigelleae*). . . . . **Ranunculaceae**
896. Plant terrestrial. 897
— Plant aquatic.—Leaves floating, peltate. Petals many. Nectaries absent. Styles and locules of the ovary many. . . . . **Nymphaeaceae**
897. Sepals or calyx-segments developing normally. 898
— Calyx cup- or saucer-shaped, margin rupturing into more or less irregular lobes. New Caledonia. (*Zygogynum*). . . . . **Winteraceae**
898. Sepals and petals either less than 6 or more than 7, rarely 6 or 7, then stigmas several. 899
— Sepals and petals 6 or 7. Stigma 1.—Shrubs. Anthers with apical pores. Embryo minute. . . . . **Ericaceae**
899. Petals imbricate, or contort, or valvate. 900
— Petals closed in bud, dropping as a cap.—Trees. Calyx apert. Tropical Africa. . . . . **Scytopetalaceae**
900. Ovary sessile or nearly so. 901
— Ovary usually long-stipitate.—Stigma 1, usually sessile. Ovules usually on the sept. Endosperm absent or nearly so. . . . **Capparaceae**
901. Anthers with apical pores or slits. 902
— Anthers with longitudinal slits. 905
902. Leaves and twigs without elastic threads (break!). Leaves alternate. Sepals and petals imbricate. Ovules axillary, or, when 2, apical, pendulous. 903
— Leaves and twigs with elastic threads. Leaves opposite. Sepals apert. Petals valvate. Ovules 2, basal, erect.—Burma to Indo-China.

Plagiopteraceae
903. Styles 3–more. Ovules numerous per locule, axillary. 904
— Style 1, shortly 3-fid. Ovules 2 per locule, collateral, apical. (*Sladeniaceae*). . . . . **Theaceae**
904. Stamens inflexed in bud. Ovary-locules numerous. Placentas not protruding into the locules. . . . . **Actinidiaceae**
— Stamens not inflexed in bud. Ovary 3–5-locular. Placentas protruding into the locules. (*Saurauiaceae*). . . . . **Actinidiaceae**
905. Bracts of sterile flowers, if any, not strongly transformed. 906
— Bracts of sterile flowers pitcher-, spoon-shaped, or saccate, brightly

coloured.—Trees or climbers. Flowers in spikes, or in racemes, or in umbels. Corolla not contort. Filaments connate at base. Ovules many per locule. Tropical America. . . . . **Marcgraviaceae**

906. Leaves opposite, rarely alternate then petals contort, filaments free or nearly so and ovules descending. 907

— Leaves alternate, stem woody, leaves rarely opposite, then stem herbaceous, petals numerous and embryo curved. 909

907. Sepals imbricate. Endosperm absent. Embryo straight. 908

— Sepals contort, at least the inner 3 when much larger than the outer 2. Endosperm copious. Embryo bent, coiled, or folded. . . . **Cistaceae**

908. Leaves with translucent-glandular stripes or dots. Ovary 2–15-locular. Ovules 1–many, when 2 not 1 ascending, 1 descending.

Guttiferae

— Leaves without such dots or stripes. Ovary 17–25-locular, locules with 1 ascending, 1 descending ovule.—Fruit umbrella-shaped. Seeds winged. Seychelles Isl. . . . . **Medusagynaceae**

909. Petals imbricate, rarely contort, then ovules ascending. 910

— Petals contort.—Flowers in panicles. Filaments more or less completely connate. Ovules 2 per locule, descending. . . . . **Meliaceae**

910. Aril absent. Endosperm scanty or absent. 911

— Aril present. Endosperm copious.—Sepals free or nearly so. Petals 5. Styles 3–more, free or connate at base only. Embryo more or less straight. . . . . **Dilleniaceae**

911. Flowers usually solitary. . . . . **Theaceae**

— Flowers in panicles.—Madagascar. (*Asteropeiaceae*, also in *Bonnetiaceae*). . . . . **Theaceae**

912. (805). Styles distinct. 913

— Stigma(s) (sub-)sessile. (*Drimys*, *Pseudowintera*). . . . . . . **Winteraceae**

913. Styles nearly completely connate.—Ovule 1 per carpel. 914

— Styles free. 916

914. Calyx valvate. Anthers with 1 slit.—Stem herbaceous. Flowers solitary. Filaments connate. Ovule 1 per carpel. Fruit dry. Endosperm present. . . . . **Malvaceae**

— Anthers with 2 slits or pores. Calyx imbricate. 915

915. Leaves translucent-glandular-punctate. Carpels warty by numerous peltate glands. Ovules 2 per carpel.—Madagascar. **Diegodendraceae**

— Leaves not punctate. Carpels not glandular-warty. Ovule 1 per carpel.—Trees or shrubs. Stipules present. Flowers in panicles. Endosperm absent. Embryo straight. . . . . **Ochnaceae**

916. Stipules absent, rarely present, then calyx imbricate and endosperm present. 917

— Stipules present. Calyx valvate. Endosperm present or not.—Woody plants. Flowers in panicles. Calyx 5-fid. Carpels 5. Seeds numerous.

Sterculiaceae

917. Herbs or undershrubs. Sepals, petals, and carpels of the same number, 6–more. Stamens twice as many. Anthers dorso-versatile. Endosperm scanty or absent.—Flowers bisexual. Ovules many.

Crassulaceae

— Sepals, petals, and carpels not of the same number, rarely so, then stamens not twice as many. Anthers usually adnate or basifix. Endosperm copious, rarely scanty or absent, then shrubs or trees.   918

918. Stipules absent, when present leaves alternate. 919

— Stipules present. Leaves opposite.—Climbing shrubs. Sepals (actually tepals) ca. 12, imbricate. Anthers petaloid, 12–25, introrse, only the outer fertile. Ovaries ca. 8, free. Styles 2-lobed. New Guinea, Queensland. . . . . **Austrobaileyaceae**

919. Calyx usually caducous. Petals 2–4, or 6–more, rarely 5 (*Ranunculaceae*), then either herbs, or twining shrubs with opposite leaves. Seeds exarillate, rarely arillate, then endosperm ruminate. 920

— Woody plants, leaves alternate, rarely erect shrubs with opposite leaves, or herbs with stipules. Calyx persistent, imbricate. Petals 5, rarely 6, then, as usual, seeds arillate, endosperm not ruminate; imbricate. . . . . **Dilleniaceae**

920. Stem herbaceous, rarely woody, but then twining and leaves opposite.—Filaments free. 921

— Stem woody, climbing or erect. Leaves alternate, undivided or lobed. 923

921. Aquatics with peltate, entire leaves. Flowers 3-merous.—Ovule 1 per carpel, parietal, pendulous. (*Brasenia*: *Cabombaceae*).

Nymphaeaceae

— Plants usually terrestrial. Leaves often incised to compound. Flowers never 3-merous. 922

922. Ovules 2–more per carpel. (*Helleboraceae*). . . . . **Ranunculaceae**

— Ovule 1 per carpel. (*Ranunculoideae*). . . . . **Ranunculaceae**

923. Carpels many, rarely 2–6, then either ovule 1, erect, or 2–more per carpel. 924

— Carpels 3–6. Ovules 2 per carpel, pendulous, descending, or patent.—Leaves not translucent-glandular-punctate. Stipules absent. Flowers unisexual, in fascicles, or in racemes, or in panicles. Mericarps drupaceous. . . . . **Menispermaceae**

924. Petals 2–6. Endosperm ruminate.—Stipules absent. . . . **Annonaceae**

— Petals 6 or more or tepals 8 or more. Endosperm absent, or if present, not ruminate. 925

925. Ovaries in a whorl, 5–20. Fruit consisting of ventrally dehiscing follicles. (*Magnoliaceae*: *Illicium*). . . . . **Illiciaceae**

— Ovaries spirally arranged. Fruit indehiscent or consisting of dorsally dehiscing follicles.............................. **Magnoliaceae**

926. (549). Stamens 1–10. 927
— Stamens 11–more... 1106

927. Ovary 1, undivided, or lobed. 928
— Ovaries 2–more, free, or connate at base and/or apex. 1097

928. Ovary 1-locular, sometimes incompletely so. 929
— Ovary completely 2–more-locular or nearly so. 967

929. Plants not obviously parasitic. Ovules not fused with each other or the ovary-wall... 930
— Mistletoe-like parasites. Ovules either fused with each other or even with the ovary-wall. .... **Loranthaceae**

930. Ovule 1... 931
— Ovules 2–more.. 941

931. Flowers unisexual. Stamens 6–10. Style simple.—Leaves undivided, alternate. Stipules absent................................... 932
— Flowers bisexual or polygamous, rarely unisexual, then either stamens 4 or 5, or leaves pinnately compound, or styles 3–5. ... 934

932. Stamens 6–10. Filaments free, or connate at base, only. 933
— Stamens 4. Filaments connate into a tube.—Indo-China, Malaya. (*Aptandraceae*: *Harmandia*). .... **Olacaceae**

933. Flowers in panicles. Stamens 6. Ovary sessile........**Simaroubaceae**
— Flowers in fascicles. Stamens 8–10. Ovary stipitate. ....**Capparaceae**

934. Flowers distinctly zygomorphic................................ 935
— Flowers actinomorphic or nearly so.—Anthers with 2 longitudinal slits.. 936

935. Leaves undivided. Sepals 5, free. Well-developed petals 3. Stamens 8.—Woody plants. .... **Polygalaceae**
— Leaves pinnately compound. Sepals connate. Well-developed petals 4 or 5. Stamens 10................................. **Leguminosae**

936. Stipules present. 937
— Stipules absent.. 938

937. Leaves opposite, undivided, tendrils absent. Stamens as many as the petals. (*Dobera, Salvadora*). .... **Salvadoraceae**
— Leaves alternate, pinnately compound. Tendrils present or not. Stamens more than petals. .... **Sapindaceae**

938. Stem woody. Petals 2–6, if 4, then stamens 1–5 or 7–10. 939
— Stem herbaceous, or woody at base only. Petals 4. Stamens 6.
Cruciferae

939. Flowers usually in racemes. Endosperm absent or nearly so.—Resiniferous (poisonous!). 940
— Flowers in racemes or umbels. Endosperm copious.—Leaves undi-

vided. Corolla valvate. Stamens as many as the petals, epipetalous.

Opiliaceae

940. Leaves opposite, paripinnate or 2-foliolate.—Female flowers in woody, many-valved cupules, formed by flattened, grooved branches. Australia. (*Blepharocaryaceae*). . . . . **Anacardiaceae**

— Leaves alternate, if opposite (*Bouea*), not compound. **Anacardiaceae**

941. Ovules 2. 942

— Ovules 3 – more. 952

942. Corolla valvate.—Woody plants. Leaves alternate, undivided. Stipules absent. 943

— Corolla imbricate. 944

943. Stamens as many as the petals, 4 or 5, alternipetalous. Ovules pendulous, apical. . . . . **Icacinaceae**

— Stamens as many as the petals, epipetalous, or less, or more. Ovules pendulous, central. (incl. *Aptandraceae*: *Aptandra*).

Olacaceae

944. Filaments free. 945

— Filaments more or less connate. 949

945. Flowers actinomorphic or nearly so. 946

— Flowers zygomorphic.—Leaves undivided. Stamens 8 – 10.

Polygalaceae

946. Stamens 3 – 5 or 8 – 10, rarely 6, then petals 3. 947

— Stamens 6. Petals 4.—Herbs or undershrubs. Leaves simple.

Cruciferae

947. Filaments inserted outside the disk, or on its edge, or between its lobes. 948

— Disk extra-staminal.—Woody plants. Leaves pinnately compound. Flowers polygamous. Stamens 5 – 8. Stigma 1. . . . . **Sapindaceae**

948. Shrubs or trees. Leaves with translucent-glandular dots, not lepidote. Flowers bisexual. Stigma 1, lobed or undivided. . . . . . . **Rutaceae**

— Woody plants. Leaves undivided, not glandular-punctate, lepidote. Flowers unisexual. Stamens 5. Stigmas 2.—Chile. (*Aextoxicaceae*).

Euphorbiaceae

949. Leaves alternate, compound, sometimes unifoliolate. 950

— Leaves opposite, undivided.—Flowers actinomorphic. Filaments 4 or 5, connate at base only. Fruit a berry. . . . . **Salvadoraceae**

950. Stipules absent. Flowers more or less actinomorphic. 951

— Stipules present. Flowers zygomorphic.—Fruit usually dehiscent.

Leguminosae

951. Stamens 5 or 6, rarely 7 – 9, all fertile. Filaments connate for most of their length. Fruit a berry. Seeds exarillate. (*Aglaia*). . . **Meliaceae**

— Stamens 10, sometimes all or the epipetalous sterile. Filaments free, or shortly connate at base only. Fruit a capsule or dry, indehiscent.

Seeds arillate. .... **Connaraceae**

952. Placenta 1, basal or central................................. 953
— Placentas 1 or more, parietal or apical...................... 956
953. Sepals usually free. Corolla imbricate or apert. Stigmas usually
     several.. 954
— Sepals connate. Corolla valvate. Stigma 1.—Plants usually woody.
     Leaves usually alternate. Placenta central. Ovules few, pendulous.
     Endosperm copious. Embryo straight. (incl. *Aptandraceae*: *Ongo-
     kea*). .... **Olacaceae**
954. Leaves alternate or radical. Embryo straight. 955
— Leaves opposite. Embryo more or less curved.—Herbs or under-
     shrubs. Endosperm copious..................... **Caryophyllaceae**
955. Plants usually woody. Leaves alternate, often scale-like. Anthers
     with longitudinal slits. Endosperm scanty or absent.... **Tamaricaceae**
— Herbs with swollen or rarely creeping rhizomes. Leaves radical,
     simple or pinnately compound. Anthers with 2 valves. Endosperm
     copious, fleshy. (*Leonticaceae*).................... **Berberidaceae**
956. Placentas 2 – more... 957
— Placentas 1.—Stipules present. Flowers usually zygomorphic.
     Stamens 9 or 10. Style 1, undivided. .... **Leguminosae**
957. Style 1, undivided.. 958
— Styles 2 – more, free, or partly connate. 959
958. Leaves simple or digitate................................... 960
— Leaves pinnately compound.—Woody plants. Flowers polygamous.
     Stamens 7 or 8, inserted within the disk. Embryo curved.
                                                        Sapindaceae
959. Flowers actinomorphic...................................... 964
— Flowers zygomorphic.—Herbs. Stamens 7 – 10. Ovary open at the
     apex. Endosperm absent. Embryo curved. .... **Resedaceae**
960. Petals usually 4. Stamens usually 6. Embryo curved.—Endosperm
     scanty or absent... 961
— Petals usually 5. Stamens usually 3 – 5 or 7 – 10. Embryo straight.—
     Plants woody. Leaves undivided. 962
961. Flowers actinomorphic or nearly so. Stamens 6, 4 longer than the
     other 2.—Leaves simple. Stipules absent. Sepals 4, free. Petals 4.
     Ovary sessile or nearly so. Placentas 2................. **Cruciferae**
— Flowers usually more or less zygomorphic. Stamens 1 – 10, when
     6 then not 4 longer than the other 2. (incl. *Cleomaceae*).
                                                        Capparaceae
962. Flowers actinomorphic. Endosperm present. 963
— Flowers zygomorphic. Endosperm absent.—Stipules absent.
     Stamens 7 – 10. Stigma 1. (incl. *Xanthophyllaceae*). .... **Polygalaceae**
963. Plants erect. Stamens not surrounded by a corona of a complicated

structure. .... **Flacourtiaceae**

— Plants climbing with tendrils. Stamens 5, inserted on the disk, surrounded by a corona of a usually complicated structure, usually composed of filamentous appendages. .... **Passifloraceae**

964. Leaves small. Stipules absent. Anthers usually extrorse. Ovules nearly basal.—Fruit a capsule. Seeds hairy. Endosperm absent, rarely present, but then placentas becoming free from the fruitwalls.
Tamaricaceae

— Leaves usually large. Stipules present or absent. Anthers usually introrse or latrorse. Ovules distinctly parietal or nearly apical.—Endosperm present. 965

965. Plants herbaceous or climbing. Ovary usually stipitate.—Corona nearly always present, outside the stamens. .... **Passifloraceae**

— Plants woody, erect. Ovary usually sessile.................... 966

966. Leaves opposite. Corona absent. Ovules 4, apical. Fruit an irregularly dehiscent capsule. Tropical Africa and Asia. (*Ctenolophon*, also in *Ctenolophonaceae* or *Olacaceae*)................. **Linaceae**

— Leaves usually alternate, rarely opposite or in whorls. Corona sometimes present. Ovules usually more than 5, usually parietal, rarely apical. Pantropical............................. **Flacourtiaceae**

967. (928). Ovule 1 per locule.................................. 968

— Ovules 2 or more per locule............................. 1008

968. Ovule erect or ascending................................. 969

— Ovule pendulous, descending or patent. 977

969. Herbs or undershrubs. Flowers actinomorphic or nearly so. Sepals 4, free. Petals 4. Stamens 6, free..................... **Cruciferae**

— Shrubs or trees, rarely herbs or undershrubs then flowers distinctly zygomorphic. Stamens more or less than 6, rarely 6 then either filaments connate, or petals 3, or sepals united. 970

970. Stamens inserted outside the disk or on its margin, 4 or 5....... 971

— Disk extrastaminal. 976

971. Leaves glandular-dotted............................... **Rutaceae**

— Leaves not glandular-dotted. 972

972. Filaments free. 973

— Filaments connate, at least at base. 975

973. Leaves simple.—Flowers actinomorphic.............. **Celastraceae**

— Leaves pinnately compound. 974

974. Leaves usually alternate. Flowers bisexual.............. **Meliaceae**

— Leaves opposite. Flowers unisexual.—Stamens 4. S. Africa. (*Ptaeroxylaceae: Ptaeroxylon*)........................ **Meliaceae**

975. Flowers bisexual. Ovules 1 or more, usually descending. . **Meliaceae**

— Flowers unisexual. Ovule 1 per locule, ascending.—Trees or shrubs with a cactus-like habit, often with spines. Madagascar. **Didieraceae**

976. Flowers unisexual or polygamous. Endosperm absent. Embryo more or less curved. . . . . **Sapindaceae**
— Flowers bisexual. Endosperm copious. Embryo straight.—Leaves compound. Stipules present. Flowers zygomorphic, 5-merous.
Melianthaceae
977. Flowers unisexual. 978
— Flowers bisexual, at least apparently so, or polygamous. 985
978. Disk extra-staminal.—Ovule apotropous. 979
— Disk intra-staminal. 980
979. Leaves pinnately compound. Flowers unisexual (staminodes sometimes very well-developed, but not functional). Petals usually imbricate. Filaments free or connate at base only. Ovary usually 3-locular. Ovule ascending. Endosperm absent. . . . . **Sapindaceae**
— Leaves simple. Flowers bisexual. Petals valvate. Filaments connate into a tube. Ovary incompletely 2-locular. Ovule pendulous. Endosperm copious. Tropical S. America, W. Africa. (*Aptandraceae: Aptandra*). . . . . **Olacaceae**
980. Flowers solitary.—Leaves pinnately compound with axillary thorns. Stamens with 2-lobed scales at base, which enclose the pistil. Fruit a schizocarp with a persistent columella. S.W. Africa. (*Neoluederitzia*). . . . . **Zygophyllaceae**
— Flowers in distinct inflorescences. 981
981. Ovule atropous.—Trees or shrubs. Leaves simple. Flowers in racemes or in panicles. Stamens 10. Styles 3 or 4. Endosperm present. (*Panda*, also in *Euphorbiaceae*). . . . . **Pandaceae**
— Ovule anatropous. 982
982. Ovule apotropous.—Resinous (often poisonous!) plants. Leaves simple or compound. Flowers in panicles. Endosperm absent.
Anacardiaceae
— Ovule epitropous. 983
983. Styles usually long and distinct. Fruit usually a 3-valved capsule. Endosperm copious. . . . . **Euphorbiaceae**
— Styles either short or stigma subsessile. Fruit a drupe or a schizocarp with indehiscent mericarps. Endosperm absent. 984
984. Flowers in cymes or in heads, these composed in panicles. Ovary 4-locular. S. Africa. (*Kirkiaceae*). . . . . **Simaroubaceae**
— Flowers in thyrses or in panicles. Ovary 2- rarely 3-locular. Asia, New Caledonia. (*Soulamea*). . . . . **Simaroubaceae**
985. Filaments more or less connate. 986
— Filaments free. 992
986. Anthers with an apical pore.—Leaves undivided. Flowers distinctly zygomorphic. . . . . **Polygalaceae**
— Anthers with 2 longitudinal slits. 987

94

987. Leaves glandular-dotted.—Leaves simple, alternate. Endosperm absent.. **Rutaceae**
— Leaves not glandular-dotted. 988
988. Herbs, sometimes woody at base, or undershrubs. Stipules present. Fruit a 5-locular schizocarp, not winged, usually awned. Temperate parts.—Leaves pinnately partite to -compound, or digitately nerved.
656
— Woody plants, rarely somewhat herbaceous, then leaves opposite or in whorls, simple, and stipules absent. Fruit a capsule, or a berry, or a drupe, rarely a schizocarp, then 2- or 3-locular and often winged. (Sub-)tropics... 989
989. Filaments connate into a tube, usually for most of their length. . 990
— Filaments connate at base, only.—Leaves simple.............. 991
990. Leaves usually pinnately compound, rarely simple. Anthers introrse. Ovules epitropous. (*Melioideae*). .... **Meliaceae**
— Leaves simple. Anthers extrorse. Ovules apotropous. (*Aptandraceae: Aptandra*). .... **Olacaceae**
991. Leaves usually opposite or in whorls. Fruit a capsule, or a schizocarp. Plants usually pubescent somewhere. 658
— Leaves alternate. Fruit a drupe. Plants entirely glaborous.—Trees or undershrubs. Stipules absent or small. Sepals imbricate, eglandular. Petals imbricate or contort. Ovary 5-locular. Endosperm copious. Tropical America, W. Africa. (*Hylocarpa, Sacoglottis*).
Houmiriaceae
992. Petals alternisepalous. Stamens episepalous, or more than the petals. 993
— Petals and stamens episepalous.—Fertile stamens 2, or 3, or 5. Styles 1 or 2. India to the Solomons. .... **Sabiaceae**
993. Styles 2 or more, free or connate at base or at the apex.—Stamens 4 or more. Anthers with longitudinal slits...................... 994
— Style 1, simple. Stigmas 1 – more............................ 997
994. Stamens 5 or 8 or 10. Fruit a drupe. Embryo straight.—Resinous (often poisonous) trees or shrubs. Ovary not or slightly lobed.
Anacardiaceae
— Stamens 4 or 6 or 8 or 10. Fruit either dry or embryo curved.... 995
995. Styles free at base, connate at the apex. (*Harrisonia*).
Simaroubaceae
— Styles or stigmas entirely free............................. 996
996. Ovary 2-, rarely 3-locular. Stigmas sessile. Seychelles to W. Pacific. (*Soulamea*)..................................... **Simaroubaceae**
— Ovary 4-locular. Styles distinct. S. Africa. (*Kirkiaceae*).
Simaroubaceae
997. Leaves simple, sometimes incised.—Leaves alternate.......... 998

95

— Leaves compound, but sometimes unifoliolate. 1004
998. Calyx valvate.—Trees. Leaves not glandular-punctate. Flowers in spikes, these arranged in panicles. Petals valvate. Ovary 4-locular. Endosperm absent. (*Poga*). . . . . **Rhizophoraceae**
— Calyx imbricate or apert. 999
999. Stem herbaceous or woody at base only.—Sepals 4, free. Petals 4, imbricate. Stamens 6, rarely less. Ovary 2-locular. Endosperm scanty or absent. Embryo large, curved. . . . . **Cruciferae**
— Stem woody.—Leaves simple. 1000
1000. Leaves glandular-punctate.—Corolla valvate or flowers fascicled. Embryo large, straight. . . . . **Rutaceae**
— · Leaves not glandular-punctate. 1001
1001. Corolla valvate.—Endosperm copious. 1002
— Corolla imbricate.—Stamens 8 – 10. 1003
1002. Inflorescence paniculate. Ovary completely divided into locules. Embryo large, curved.—Disk small, annular. Stamens 4 or 5.

Icacinaceae
— Inflorescence usually fasciculate. Ovary not completely divided into locules, 1-locular near the apex. Embryo small, straight, on top of the endosperm. . . . . **Olacaceae**
1003. Flowers in a panicle. Disk large, cushion-shaped. Fruit a drupe. Endosperm absent or nearly so. Embryo large, straight, or nearly so. (also in *Linaceae* or *Simaroubaceae*). . . . . **Irvingiaceae**
— Flowers in a raceme. Disk little developed. Fruit dry, indehiscent. Endosperm copious. Embryo small, in the centre of the endosperm. America. . . . . **Cyrillaceae**
1004. Leaves not translucent-glandular-punctate. 1005
— Leaves translucent-glandular-punctate. . . . . **Rutaceae**
1005. Stipules absent, or leaves with 2 sub-basal spines. 1006
— Stipules present, rarely absent, then with 1 axillary spine (*Balanitaceae*).—Flowers bisexual. Stamens 10, often appendiculate at base. Disk intra-staminal. Stigma usually lobed. Ovary 5-, or 10-, 12-locular. Fruit usually a schizocarp. Embryo straight. **Zygophyllaceae**
1006. Disk intra-staminal. Stamens 8 – 10. 1007
— Disk extra-staminal. Stamens 5 – 8.—Flowers polygamous. Ovary 2- or 3-locular. Embryo curved. . . . . **Sapindaceae**
1007. Flowers bisexual. Stamens with 2-lobed appendages at base. Stigma 4- or 5-lobed. Embryo curved. Tropical Africa to Australia. (*Harrisonia*). . . . . **Simaroubaceae**
— Flowers polygamous. Stamens unappendaged. Stigmas 4 or 5, filiform. Embryo straight. Mexico. (*Cyrtocarpa*). . . . . **Anacardiaceae**
1008. (967). Ovules 2 per locule. 1009
— Ovules 3 or more per locule. 1062

1009. Ovules erect or ascending or patent or one ascending and the other descending. . . . .—. 1010
— Ovules pendulous or descending. 1026
1010. Filaments more or less connate. 1011
— Filaments free. 1015
1011. Disk extra-staminal. 1012
— Disk intra-staminal. 1013
1012. Shrubs or trees. Leaves alternate, usually pinnately compound. Petals imbricate. Endosperm absent. (*Dodonaeoideae*). **Sapindaceae**
— Herbs or undershrubs. Leaves opposite or in whorls, simple. Petals contort. Endosperm present.—Chile, S. Brasil. (*Vivianiaceae*).
Geraniaceae
1013. Leaves glandular-punctate. Ovary deeply lobed, rarely terete, but then leaves 1–3-foliolate. . . . . **Rutaceae**
— Leaves pinnately compound or simple, rarely 1–3-foliolate, but then not glandular-punctate. Ovary terete or only slightly lobed. 1014
1014. Flowers zygomorphic. Stigma simple, punctiform. Capsule inflated, membranous, loculicide.—S. Africa. (*Aitoniaceae*). . . . . . . **Meliaceae**
— Flowers actinomorphic. Stigma not both simple and punctiform. Fruits otherwise. . . . . **Meliaceae**
1015. Fertile stamens as many as the sepals, or more, 3–10. 1016
— Fertile stamens less than the sepals, 2 or 3.—Leaves usually opposite. Ovary 3-locular. Style 1. (also in *Celastraceae*).
Hippocrateaceae
1016. Flowers bisexual. Sepals 4. Petals 4. Stamens 6, unequally long. Ovary 2-locular.—Herbs or undershrubs. . . . . **Cruciferae**
— Stamens 3–5 or 7–10, rarely 6, but then either petals 3 or 6, or flowers unisexual or polygamous or stamens equally long. 1017
1017. Disk extrastaminal.—Flowers unisexual or polygamous. 1018
— Stamens inserted outside the disk or on it (near the margin). . . 1020
1018. Leaves alternate. . . . . **Sapindaceae**
— Leaves opposite. 1019
1019. Flowers actinomorphic. Stamens twice as many as sepals, rarely as many and episepalous. Ovary 2-locular. Stigmas 2. . . . . **Aceraceae**
— Flowers zygomorphic. Stamens more than sepals, but less than twice as many, rarely as many but then alternisepalous. Ovary 3-locular. Stigma 1.—Leaves compound. . . . . **Hippocastaneaceae**
1020. Stamens as many as petals and epipetalous.—Leaves alternate. Stipules present. Corolla valvate. . . . . **Vitaceae**
— Stamens as many as petals, alternipetalous, or more. 1021
1021. Stamens as many as petals, or petals 4 and stamens 5 or 6.—Leaves simple, not glandular-punctate. 1022
— Stamens twice as many as petals, rarely as many, but then leaves

glandular-punctate.. 1023

1022. Stamens as many as petals. Ovules usually collateral, basal, rarely superposed and one ascending and the other descending (*Maytenus*), but then stipules present and fruit a capsule.—Not wild in New Zealand. .... **Celastraceae**

— Stipules absent. Petals 4. Stamens 4–6. Ovules superposed, one ascending, the other descending. Fruit a berry.—New Zealand. (*Aristotelia*). .... **Elaeocarpaceae**

1023. Stipules present. 1024

— Stipules absent.. 1025

1024. Leaves simple. Sepals valvate. Anthers with terminal pores.

Elaeocarpaceae

— Leaves simple or 3-partite. Sepals imbricate. Anthers with 2 longitudinal slits. (*Fagonia*)........................ **Zygophyllaceae**

1025. Leaves glandular-punctate............................. **Rutaceae**

— Leaves not glandular-punctate.—Leaves opposite. Flowers unisexual or polygamous. Stigmas 2. .... **Aceraceae**

1026. (1009). Stipules present. 1027

— Stipules absent.. 1043

1027. Leaves opposite.. 1028

— Leaves alternate, rarely (*Geraniaceae*) opposite, but then filaments connate at base and styles or stigmas 2–5................... 1032

1028. Stamens 5 or less... 1029

— Stamens 8–10. 1030

1029. Stamens 2 or 3. (also in *Celastraceae*)............. **Hippocrateaceae**

— Stamens 4 or 5....................................... **Celastraceae**

1030. Filaments free. Seeds not winged. 1031

— Filaments connate. Seeds winged.—Large trees. Sepals and petals valvate. Fruit a capsule. W. Africa. (*Anopyxis*)..... **Rhizophoraceae**

1031. Woody plants. Leaves simple. Sepals and petals valvate. Filaments inappendiculate. Fruit a berry. New Guinea. (*Sericolea*).

Elaeocarpaceae

— Herbs or shrubs. Leaves pinnately compound, rarely simple, plant then a succulent annual with valvate sepals and apert, trifid petals (*Augea*), otherwise sepals and petals imbricate. Filaments appendiculate. Fruit a capsule or a schizocarp........... **Zygophyllaceae**

1032. Flowers bisexual. Fruit dry. 1033

— Flowers unisexual, rarely bisexual or polygamous, then fruit a drupe and stamens usually free. 1039

1033. Styles or stigmas 2–5................................... 1034

— Style 1. Stigma 1, entire or lobed. 1036

1034. Herbs or woody perennials. Stamens 5–10. Filaments connate at base.—Disk extra-staminal. (*Geranieae*). .... **Geraniaceae**

98

— Stem woody. Stamens 10, free............................. 1035
1035. Calyx valvate. Disk intra-staminal...................... **Tiliaceae**
— Calyx imbricate. Disk extra-staminal.—Sepals 3. Madagascar. (*Leptolaena*). .... **Sarcolaenaceae**
1036. Stamens 6, free. .... **Capparaceae**
— Stamens 8–10, free or connate. ....: 1037
1037. Stipules free. Sepals usually 5, subequal. 1038
— Stipules intra-petiolarily connate. Sepals 8–10, very unequal.— Stamens 10, filaments free. New Caledonia....... **Strasburgeriaceae**
1038. Filaments more or less connate. .... **Meliaceae**
— Filaments free.—Flowers solitary, terminal. Petals bright yellow. Somalia. (*Kelleronia*)........................... **Zygophyllaceae**
1039. Leaves undivided or unifoliolate, then flowers 5-merous. 1040
— Leaves pinnately compound.—Flowers 3- or (*Garuga*) 5-merous.

Burseraceae
1040. Sepals 5, imbricate. Fruit a drupe. Endosperm absent.—Stamens 5. Stigmas 2 or 3. .... **Dichapetalaceae**
— Fruit a capsule, rarely a berry or a drupe, but then calyx valvate. Endosperm usually copious................................ 1041
1041. Leaves simple, pedicel not articulated, stipels absent. Disk extra-staminal... 1042
— Leaves unifoliolate, pedicel articulated, stipel 1, early fugacious (scar!). Disk intra-staminal.—Flowers unisexual. Ovule and seed with a caruncle. Ovules collateral. Fruit a capsule. Nigeria to Congo. .... **Lepidobotryaceae**
1042. Flowers unisexual. Ovules collateral. Caruncle present on ovules and seeds. (*Phyllanthoideae*, incl. *Centroplacus*, generally included in *Pandaceae*)................................. **Euphorbiaceae**
— Flowers bisexual or polygamous. Ovules more or less serial. Caruncle absent.—Fruit a drupe..................... **Elaeocarpaceae**
1043. (1026). Stamens less than petals and alternipetalous........... 1044
— Stamens as many as petals or more, or less and then epipetalous.

1045
1044. Petals 4. Stamens 2................................. **Cruciferae**
— Petals 5. Stamens 3, rarely 2 or 4. (also in *Celastraceae*).

Hippocrateaceae
1045. Stamens as many as petals and epipetalous (some stamens sometimes sterile) or less (and then epipetalous). 1046
— Stamens as many as petals and alternipetalous, or more. 1048
1046. Flowers unisexual. Petals alternating with the sepals or calyx-lobes. Stamens in the male flowers all fertile.—Leaves pinnately compound. (*Picramnia*). .... **Simaroubaceae**
— Flowers bisexual or polygamous. Petals opposed to the sepals.

Stamens *either* all fertile and then leaves simple, *or* stamens 5, only 2 or 3 of which fertile, then leaves simple or pinnately compound.. 1047

1047. Fertile stamens 2 or 3. Staminodes 3 or 2.—Leaves simple or pinnately compound. (*Meliosma, Ophiocaryon*, also placed in *Meliosmaceae*)... **Sabiaceae**

— Fertile stamens 5. (*Sabia*)............................... **Sabiaceae**

1048. Filaments connate at least at base.—Leaves pinnately compound or undivided, rarely 1–3-foliolate then not glandular-punctate. ... 1049

— Stamens free or inserted on the disk, rarely filaments connate at base, then either leaves 1–3-foliolate and glandular-punctate, or ovary deeply lobed. 1050

1049. Bark of twigs and petioles with a light-coloured, wavy, sclerenchymatic band and with resin ducts between this and the wood cylinder. Filaments connate at base only. (*Canarium, Scutinanthe*).

Burseraceae

— Bark of twigs and petioles without such a band and not resinous. Filaments connate into a tube for most of their length. (*Melioideae*).

Meliaceae

1050. Stamens 6, rarely 4, then, as usual, herbs with non-glandular-punctate leaves. Sepals 4, free. Petals 4................ **Cruciferae**

— Stamens as many as or (nearly) twice as many as the petals. Shrubs or trees, rarely herbs, then either leaves glandular punctate, or flowers 5-merous. 1051

1051. Stamens 3 or 4.—Leaves alternate, simple, translucent-glandular-punctate. Peduncle adnate to the petiole of its bract. Flowers bisexual, solitary, or in cymes, axillary. Petals 3 or 4, imbricate. Style 1. Stigmas 3 or 4. Schizocarp dehiscing into 3 or 4 drupelets, columella persistent. Endosperm fleshy. Embryo horse-shoe-shaped.

Cneoraceae

— Stamens 5–10, if 3 or 4 plant not as above.................. 1052

1052. Leaves opposite, not punctate. 1053

— Leaves alternate, rarely opposite, but then translucent-glandular-punctate.. 1056

1053. Flowers unisexual or polygamous. Fruit dehiscent into 2 samaras. Endosperm absent. Stigmas 2......................... **Aceraceae**

— Flowers bisexual. Style simple with 1 stigma or with 2 branches with 1 stigma each. Endosperm present. 1054

1054. Stamens 5. Fruit a drupe.—Corolla imbricate......... **Celastraceae**

— Stamens 10. Fruit a capsule. 1055

1055. Shrubs. Corolla induplicative-valvate. Australia..... **Tremandraceae**

— Trees. Corolla imbricate. Tropical Africa and Asia. (*Ctenolophon*, also in *Olacaceae* or *Ctenolophonaceae*)................. **Linaceae**

100

1056. Leaves not translucent-glandular-punctate, alternate.......... 1057
— Leaves translucent-glandular-punctate.—Bark resinous. Fruit sometimes a drupe, then either leaves opposite, or endosperm present.

Rutaceae

1057. Bark resinous (also in the twigs).—Leaves usually compound. Stigma 1. Fruit drupaceous, but sometimes dehiscent. Endosperm absent. .... **Burseraceae**
— Bark not resinous. 1058
1058. Leaves undivided. Endosperm present...................... 1059
— Leaves compound. Endosperm absent.—Stigma 1............. 1060
1059. Fruit a capsule. Stamens 5. Stigmas 2 or 3. .... **Cyrillaceae**
— Fruit a drupe or dry and indehiscent. Stamens 4 or 5. Stigmas 1 or 4
or 5.. 1061
1060. Petals 4 or 5. Stamens 8–10. Disk intra-staminal. Fruit a capsule or
a berry.. **Meliaceae**
— Stamens 5–8, less than twice as many as petals. Disk extra-
staminal. Fruit a capsule. .... **Sapindaceae**
1061. Erect woody plants, or lianas rarely with tendrils (*Iodeae*), then
stigma simple to indistinctly lobed, or twining herbs. Base of leaves
without warty fields. Ovary 1-locular, sometimes also 2 abortive
locules present. Fruit usually drupaceous, not winged.... **Icacinaceae**
— Lianas with tendrils. Base of leaves with warty fields. Ovary (4- or)
5-lobed. Fruit indehiscent, winged.—S.E. Asia, W. Pacific.

Lophopyxidaceae

1062. (1008). Ovules basal, axillary, or apical..................... 1063
— Ovules parietal.—Endosperm absent. Embryo usually curved. . 1064
1063. Ovules axillary or basal. 1066
— Ovules apical.—Leaves alternate, not translucent-glandular-
punctate. Stipules absent. Stamens 5. Styles 2 or 3. Ovules 3, pendulous. America. .... **Cyrillaceae**
1064. Stamens 6–10, rarely 4. Ovary usually stipitate.—Shrubs or trees.
Stigma 1. Fruit juicy, berry-like..................... **Capparaceae**
— Stamens 4–6. Ovary usually sessile.—Herbs or shrubs or treelets.

1065

1065. Herbs or undershrubs. Stipules absent. Ovary 2-, rarely 3- or
4-locular. Fruit dry, usually dehiscent.—Sepals 4, free. Petals 4,
imbricate. Stamens 6, unequal, rarely less. Stigmas 1 or 2.

Cruciferae

— Shrubs or treelets. Stipules present. Ovary 4–7-locular. Fruit a
drupe.—Sepals, petals and stamens equal in number, 4–6. Sepals
imbricate or valvate. Petals imbricate. Stigma 4–7-lobed. (*Brexia*,
Brexiaceae). .... **Saxifragaceae**
1066. Style 1, simple. 1073

— Styles 2–5 and free or stigmas 2–5 and sessile. 1067
1067. Leaves opposite or in whorls. 1068
— Leaves alternate or radical. 1071
1068. Stipules absent. 1069
— Stipules present.—Stem woody. 1070
1069. Leaves simple. Stamens 9 in 3 bundles. Styles 3. . . . . **Guttiferae**
— Leaves 3-foliolate, seemingly in whorls of 6 leaflets. Stamens numerous, free. Styles 2. (*Bauera, Baueraceae*). . . . . **Saxifragaceae**
1070. Stamens 8 or 10. Styles 2 or 3.—Shrubs or trees. . . . . . . **Cunoniaceae**
— Stamens 5. Styles 3. . . . . **Staphyleaceae**
1071. Stem herbaceous. Stipules absent. Stamens 4 or 8 or 10. 1072
— Stem woody. Stipules small, early fugaceous. Stamens 5.—Brazil, Guianas. (*Goupiaceae*). . . . . **Celastraceae**
1072. Leaves alternate, undivided. Ovary deeply lobed. Styles 3 or 4.
Crassulaceae
— Leaves radical, lobed. Ovary weakly lobed. Styles or stigmas 4 or 5. . . . . **Saxifragaceae**
1073. Stamens as many as petals or less. 1074
— Stamens more than petals. 1083
1074. Anthers dehiscing with 2 longitudinal slits or with 1 transverse slit.
1076
— Anthers dehiscing with 2 apical pores or with 1 longitudinal slit.— Shrubs. Leaves alternate, undivided. Stipules absent. Flowers actinomorphic. Ovary 5-locular. Style 1. 1075
1075. Sepals usually connate, sometimes absent or free. Anthers dehiscing with 2 apical slits. . . . . **Ericaceae**
— Sepals entirely free and imbricate. Anthers dehiscing with 1 longitudinal slit. . . . . **Epacridaceae**
1076. Flowers zygomorphic. 1077
— Flowers actinomorphic. 1078
1077. Leaves opposite, undivided. Petals 5. Stamens connate. Ovary 3-locular. . . . . **Trigoniaceae**
— Leaves alternate, pinnately compound. Petals 4. Stamens free. Ovary 4-locular.—Stipules present. Sepals 5. Fruit a capsule.
Melianthaceae
1078. Leaves pinnately compound, rarely simple, then translucent-glandular-punctate. 1079
— Leaves simple, not partite, not translucent-glandular-punctate, usually opposite. 1082
1079. Stipules present, sometimes early fugacious (!). 1080
— Stipules absent. 1081
1080. Woody plants. Ovary 3-locular. . . . . **Staphyleaceae**
— Herbs. Ovary 4- or 5-locular. (*Tetradiclis, Tribulus*). **Zygophyllaceae**

102

1081. Trees. Leaves not translucent-glandular-punctate. Stamens inserted on the upper margin of a cushion-shaped or columnar disk.—Leaves alternate. Flowers in racemes. Stigma 1, discoid. Ovary 4- or 5-locular. . . . . **Meliaceae**
— Woody plants. Leaves translucent-glandular-punctate. Stamens usually inserted at the base of a cup-shaped disk.—Ovary 3-, or 5-more-locular. (incl. *Flindersiaceae*). . . . . **Rutaceae**
1082. Stamens as many as petals, 4 or 5, inserted on or outside the disk. Anthers usually introrse. Endosperm usually present. . . **Celastraceae**
— Stamens less than petals, 3 or rarely 2 or 4, inserted on or inside the disk, very rarely as many as the petals, 5, and inserted within the disk. Anthers extrorse. Endosperm absent. . . . . **Hippocrateaceae**
1083. (1073). Filaments free. 1084
— Filaments more or less connate. 1094
1084. Ovary sessile, rarely stipitate, but then either ovary deeply lobed, or leaves compound. 1085
— Ovary usually stipitate, undivided.—Woody plants. Leaves alternate, simple. Fruit a berry. 1092
1085. Leaves compound and stamens 5–8, or leaves simple, then not translucent-glandular-punctate and stamens up to 10. 1086
— Leaves compound, rarely simple, then translucent-glandular-punctate. Stamens 8–10.—Stamens inserted outside the disk or on its margin. Anthers with 2 longitudinal slits. 1093
1086. Stipules present (scars).—Leaves simple. Calyx valvate. 1087
— Stipules absent.—Leaves simple or compound. 1088
1087. Inflorescence usually elongate. Corolla valvate. Anthers dehiscing apically. . . . . **Elaeocarpaceae**
— Flowers in axillary fascicles. Corolla apert or slightly imbricate. Anther dehiscing longitudinally. (*Gynotroches*). . . . . **Rhizophoraceae**
1088. Leaves compound.—Stamens 5–8. Disk extra-staminal. Anthers with longitudinal slits. Ovary 3-locular. . . . . **Sapindaceae**
— Leaves simple. 1089
1089. Herbs. 1090
— Woody plants. 1091
1090. Autotrophic plants with well-developed, green leaves. Anthers incurved in bud, with 2 apical pores or tubules. . . . . **Pyrolaceae**
— Non-green saprophytes without well-developed leaves. Anthers erect in bud, thecae with a common slit, or with 2 longitudinal slits.
Monotropaceae
1091. Shrubs. Stamens 6–10, inserted on the margin of the disk or outside the disk. Anthers dehiscing by 2 pores or slits. . . . . . . **Ericaceae**
— Small trees. Stamens 10. Disk extra-staminal. Anthers dehiscing longitudinally.—Ovary 5-locular. (*Greyaceae*). . . . . . . **Melianthaceae**

1092. Petals 2–4, free. Stigma sessile.—Endosperm absent. . . **Capparaceae**
— Petals 5, coherent at base. Style 1. Stigma 3–5 lobed. Tropical W. Africa. (*Pentadiplandraceae*, sometimes in *Celastraceae*). **Capparaceae**
1093. Leaves translucent-glandular-punctate. Stipules absent. Stamens usually inappendiculate. (incl. *Flindersiaceae*). . . . . **Rutaceae**
— Leaves not translucent-glandular-punctate. Stipules present. Stamens usually appendiculate.—Calyx and corolla imbricate.

Zygophyllaceae

1094. (1083). Leaves opposite or in whorls. 1095
— Leaves alternate. 1096
1095. Leaves translucent-glandular-punctate. Stipules absent. Flowers actinomorphic. Stamens 8–10. Ovary 4- or 5-locular. . . . . . **Rutaceae**
— Leaves not glandular-punctate. Stipules present. Flowers zygomorphic. Stamens 6. Ovary 3-locular.—Leaves undivided. Petals 5.

Trigoniaceae

1096. Leaves undivided. Flowers usually zygomorphic. Embryo curved.— Ovary stipitate. . . . . **Capparaceae**
— Leaves pinnately compound, rarely simple, then, as usual, flowers actinomorphic, ovary sessile to immersed in the disk, and embryo straight.—Stipules absent. Ovary 2–6-locular. . . . . **Meliaceae**
1097. (927). Styles or stigmas connate. 1098
— Styles and stigmas completely free. 1100
1098. Leaves translucent-glandular-punctate. . . . . **Rutaceae**
— Leaves not translucent-glandular-punctate.—Shrubs or trees. Filaments free. Ovule 1 per carpel. 1099
1099. Disk extra-staminal.—Leaves usually paripinnate, sometimes imparipinnate or simple. . . . . **Sapindaceae**
— Disk intra-staminal.—Leaves either imparipinnate or simple.

Simaroubaceae

1100. Ovules 1 or 2 per carpel.—Shrubs or trees. 1101
— Ovules numerous, rarely 1 or 2 per carpel, then plant a herb or an undershrub (*Crassulaceae*). 1103
1101. Leaves simple, undivided, not translucent-glandular-punctate. . . 1102
— Leaves compound, if simple translucent-glandular-punctate. . . . 1104
1102. Ovule 1 per carpel, more or less apical, or 2. . . . . **Simaroubaceae**
— Ovule 1 per carpel and basal.—Stamens 8–10. . . . . . . **Anacardiaceae**
1103. Herbs or undershrubs. Leaves simple. Fruit a capsule. **Crassulaceae**
— Lianas with palmately compound leaves or trees with pinnately compound leaves. Fruit composed of berries.—Sepals and petals 3. Stamens 6. . . . . **Lardizabalaceae**
1104. Ovule 1 per carpel, more or less apical. (incl. *Kirkiaceae*).

Simaroubaceae

— Ovules 2 per carpel. 1105

104

1105. Leaves translucent-glandular-punctate. Stamens 3–5.—Endosperm present... **Rutaceae**
— Leaves rarely translucent-glandular-punctate, then endosperm absent. Stamens 10, sometimes 5 staminodial............ **Connaraceae**
1106. (926). Ovary 1, undivided or lobed......................... 1113
— Ovaries 2 or more, free or only connate at base. 1107
1107. Styles entirely free. 1108
— Styles connate, at least at the base or at the apex............. 1110
1108. Stipules absent. Flowers actinomorphic. 1109
— Stipules present. Flowers zygomorphic.—Flowers in spikes or in racemes. Sepals 5, connate at base. Anthers introrse. Carpels 5 or 6. Ovules 1–3 per carpel........................... **Resedaceae**
1109. Ovule 1 per carpel.—Aquatics. Flowers solitary. Sepals free, numerous. Anthers extrorse. Carpels 9–17. .... **Nympheaceae**
— Ovules several or many per carpel.—Leaves simple. 1112
1110. Leaves undivided. Stipules present.—Disk intra-staminal. Anthers adnate. .... **Ochnaceae**
— Leaves absent or pinnately compound. Stipules absent. .....:.. 1111
1111. Sepals free. Disk extra-staminal. Ovaries 2 or 3........ **Sapindaceae**
— Sepals connate at base. Disk intra-staminal. Ovaries 5 or 6.—Subtropical and tropical America. (*Castela, Quassia*). ... **Simaroubaceae**
1112. Shrubs or trees. Sepals 3. Anthers adnate. Fruit a berry.
Annonaceae
— Herbs or undershrubs. Sepals 6 or more. Anthers versatile. Fruit a capsule.. **Crassulaceae**
1113. Ovary 1-locular, sometimes incompletely so. 1114
— Ovary 2–more-locular, sometimes nearly so................. 1124
1114. Ovule 1. 1115
— Ovules 2–more... 1117
1115. Leaves opposite.—Flowers polygamous, solitary or in fascicles. Petals 4. Ovary sessile. .... **Guttiferae**
— Leaves alternate. 1116
1116. Flowers in fascicles. Petals 2–4. Stigma sessile.—Flowers unisexual. Ovary stipitate............................... **Capparaceae**
— Flowers in panicles. Petals 3 or 5 or 6. Style well-developed.
Anacardiaceae
1117. Leaves opposite.. 1118
— Leaves alternate. 1119
1118. Style simple. Embryo curved.—Petals 5 and contort, or 3. Calyx imbricate. Ovules usually atropous. Fruit a capsule......... **Cistaceae**
— Styles or style-branches 2–5. Embryo straight.—Leaves undivided, opposite. Stipules absent. Flowers actinomorphic. .... **Guttiferae**
1119. Flowers distinctly zygomorphic. Ovary open at apex.—Herbs.

105

Leaves simple, undivided, or pinnately partite. Flowers in spikes or in racemes. Styles 3–6. Endosperm absent or nearly so. **Resedaceae**

— Flowers actinomorphic or slightly zygomorphic. Ovary closed at apex. 1120

1120. Flowers actinomorphic. Ovary and fruit usually not stipitate, if so plant woody, branches stellately pubescent, and leaves linear, sinuately lobulate.—Woody plants. Endosperm usually present, if scanty leaves linear or scale-like. 1121

— Flowers slightly zygomorphic. Ovary and fruit usually stipitate, if sessile plants glandular annuals (*Cristatella*).—Either herbs with 3-foliolate or palmately compound leaves, or shrubs. Style 1, simple. Endosperm scanty to absent. (incl. *Cleomaceae*). . . . . . . **Capparaceae**

1121. Leaves often linear or scale-like. Stipules absent. Endosperm scanty. —Flowers solitary. Calyx imbricate. Anthers extrorse. Fruit a capsule, with placenta separating from the wall. Seeds hairy. . . . . . . **Tamaricaceae**

— Leaves normally developed. Stipules present, often soon caducous. Endosperm present. 1122

1122. Disk appendiculate. 1123

— Disk inappendiculate.—Flowers in racemes. Sepals 4 or 5, valvate, free. Stamens more than 10. Style simple. Stigma small. Australia (*Nettoa*, ? once found) or New Caledonia (*Oceanopapaver*, also in *Capparaceae*, doubtfully included here). . . . . **Tiliaceae**

1123. Calyx imbricate or valvate. Ovules anatropous. Embryo straight. (incl. *Prockieae*, also in *Tiliaceae*). . . . . **Flacourtiaceae**

— Calyx imbricate. Ovules usually atropous. Embryo curved.—Petals 5, contort, or 3. Style simple. Fruit a capsule. . . . . **Cistaceae**

1124. (1113). Ovules 1 or 2 per locule. 1125

— Ovules 3 or more per locule. 1139

1125. Flowers bisexual, at least apparently so, or polygamous, rarely unisexual, then stipules and endosperm absent. 1126

— Flowers unisexual.—Stipules usually present (early fugacious!). Ovules anatropous, pendulous, axillary, usually with a caruncle. Endosperm present. Embryo straight. . . . . **Euphorbiaceae**

1126. Calyx valvate. 1127

— Calyx imbricate or apert. 1129

1127. Petals entire.—Stipules present. Endosperm present. . . . . **Tiliaceae**

— Petals dentate or fimbriate. 1128

1128. Flowers in racemes or corolla imbricate. . . . . **Elaeocarpaceae**

— Flowers in simple or compound cymes. Corolla valvate. (*Anopyxis*, *Crossostylis*). . . . . **Rhizophoraceae**

1129. Corolla imbricate, or contort, rarely valvate, then calyx divided up to halfway. 1130

— Corolla valvate.—Calyx slightly lobed or dentate. . . . . **Olacaceae**

1130. Leaves alternate, rarely opposite, then either compound or stipules present.. 1131
— Leaves opposite, undivided.—Leaves often translucent-glandular-punctate or -striped. Stipules absent. Endosperm absent. **Guttiferae**
1131. Style 1. 1132
— Styles 3.—Resinous (often poisonous!), usually woody plants. Leaves pinnately compound, not translucent-glandular.

Anacardiaceae

1132. Filaments free. 1133
— Filaments connate, at least at base. 1137
1133. Leaves not translucent-glandular-punctate.................... 1134
— Leaves translucent-glandular-punctate.—Shrubs or trees. Leaves 1- or 3-foliolate or undivided. Stipules absent. Disk intra-staminal.

Rutaceae

1134. Trees or woody lianas.—Disk extra-staminal. 1135
— Shrubs or herbs. 1136
1135. Leaves pinnately compound. Stipules absent. Ovule 1 per locule. Endosperm absent................................. **Sapindaceae**
— Leaves simple. Stipules present. Ovules 2 per locule. Endosperm present.. **Sarcolaenaceae**
1136. Herbs. Basal leaves bipinnatipartite. Stipules absent. Flowers 4-merous, in panicles. Petals imbricate. (*Megacarpaea*)..... **Cruciferae**
— Shrubs. Leaves entire or apically trifid. Stipules present. Flowers 5-merous, in cincinni. Petals valvate. (*Nitraria*)....... **Zygophyllaceae**
1137. Leaves simple. Filaments connate at base only. 1138
— Leaves pinnately compound. Filaments connate into a tube.

Meliaceae

1138. Shrubs or trees. Flowers in cymes or in panicles. Stigma 1. Fruit a drupe. Endosperm present. Embryo straight.......... **Houmiriaceae**
— Herbs or undershrubs. Flowers solitary or in umbels. Stigmas 5. Fruit a capsule. Endosperm absent. Embryo curved. (*Monsonia, Sarcocaulon*)................................. **Geraniaceae**
1139. (1124). Ovary ± distinctly stipitate.—Shrubs or trees. Leaves alternate. Stigma 1. Fruit a berry. 1140
— Ovary ± sessile.—Either endosperm present, or embryo straight.

1141

1140. Petals 2–4, free. Stigma sessile.—Endosperm absent. Embryo curved. .... **Capparaceae**
— Petals 5, coherent at base. Style 1. Stigma 3–5-lobed.—Stamens 11–13. Tropical W. Africa. (*Pentadiplandraceae*, sometimes in *Celastraceae*). .... **Capparaceae**
1141. Calyx valvate.—Stipules present. Endosperm present......... 1143
— Calyx imbricate or apert. 1142

1155. Herbs or undershrubs, rarely shrubs.—Leaves simple, usually opposite. Petals usually minute. Stigmas 2 or 3....... **Caryophyllaceae**
— Woody plants.. 1156
1156. Trees. Leaves pinnately compound. Stigmas 2 or 3.... **Staphyleaceae**
— Woody plants. Leaves simple, opposite. Stigmas 5.—Stamens 5, epipetalous. .... **Rhamnaceae**
1157. Flowers actinomorphic, rarely slightly zygomorphic, then not papilionate.. 1158
— Flowers usually zygomorphic and papilionate, when actinomorphic stipules present, as usual, and ovule parietal.—Stamens 8–10.
Leguminosae
1158. Stipules present, sometimes minute, and/or early fugacious..... 1159
— Stipules absent.—Shrubs or trees.......................... 1163
1159. Corolla valvate... 1160
— Corolla imbricate or apert............................... 1161
1160. Plants usually herbaceous. Leaves lobed, or partite, or compound. Endosperm absent.—N. America. (*Gillenia*). .... **Rosaceae**
— Trees. Leaves undivided. Endosperm copious.—Flowers solitary or in fascicles. Sepals valvate, calyptrate. Stamens 8–10. Anthers quadrangular. Ovule erect. Tropical Africa. (*Hua*, also in *Sterculiaceae* or *Styracaceae*). .... **Huaceae**
1161. Leaves undivided. Corolla imbricate. Ovule erect, basal. 1162
— Leaves lobed. Flowers in capitules. Corolla apert. Ovule pendulous.—Endosperm present. (*Platanus*).............. **Platanaceae**
1162. Flowers cymose or solitary. Stamens 4 or 5. Style terminal.
Celastraceae
— Flowers in a terminal panicle. Stamens 3–10. Style gynobasic.—Madagascar, Tropical America. (*Hirtella*). .... **Chrysobalanaceae**
1163. Staminodes petaloid.—Trees without resin. Leaves alternate, undivided. Flowers in panicles. Fertile stamens epipetalous. Ovule pendulous, anatropous, apotropous. Endosperm present.
Corynocarpaceae
— Staminodes not petaloid or absent......................... 1164
1164. Resiniferous plants. Bark not silky fibrous inside. Flowers usually in panicles. Ovule with dorsal raphe, usually erect, micropyle downwards.—Endosperm absent or nearly so. .... **Anacardiaceae**
— Plants without resin. Bark inside with tough silky fibres. Flowers in spikes, or in racemes, or in capitules, or in umbels, or solitary, rarely in panicles. Ovule with ventral raphe, pendulous.—Leaves simple. .... **Thymelaeaceae**
1165. (1153). Ovules 2. 1166
— Ovules 3 or more. 1177
1166. Leaves undivided or lobed. 1167

109

— Stipules early fugacious. Ovules erect, 6.—Flowers in dense, sub-globose umbels. Tepals 10–14, valvate. Disk-glands alternating with the 5–7 stamens, epitepalous. Capsule short-hairy. S.E. Asia.

Dipentodontaceae

1180. Calyx valvate.—Embryo straight. 1181
— Calyx imbricate.—Leaves opposite. Endosperm present. 1182
1181. Stigma 1. Endosperm absent. . . . . **Lythraceae**
— Stigmas 2 or 3. Endosperm scanty.—Woody plants. Stamens 5, epipetalous. Ovary incompletely locular, ovules 4–6. . . **Rhamnaceae**
1182. Herbs or undershrubs. Stamens 1–5. Stigmas 2 or 3, rarely 1, elon-gated.—Embryo more or less curved. . . . . **Caryophyllaceae**
— Stem woody. Stamens 5. Stigma 1, peltate. . . . . **Celastraceae**
1183. Placenta 1, parietal. 1184
— Placentas 2–more. 1186
1184. Calyx valvate or descendingly imbricate (*i.e.* the odd sepal above).
1185
— Calyx ascendingly imbricate (*i.e.* the odd sepal below), rarely closed, or apert, or valvate, then leaves simple and entire, or 2-lobed or -partite, or, as usual, compound.—Stipules present.

Leguminosae

1185. Stem woody. Leaves alternate, dentate, or 3–9-lobed. Stipules present. . . . . **Rosaceae**
— Stem herbaceous. Leaves opposite, entire. Stipules absent.—Ovary occasionally 2-locular with 1 empty locule. . . . . **Lythraceae**
1186. Leaves simple. Anthers with 2 longitudinal slits or apical pores. 1187
— Leaves pinnately compound. Anthers with 1 longitudinal slit.—Trees. Flowers zygomorphic. Fertile stamens 5, epipetalous. Ovules many. Endosperm absent. . . . . **Moringaceae**
1187. Stamens as many as the petals, alternipetalous. 1188
— Stamens more than petals, rarely as many, then epipetalous. . . 1191
1188. Tendrils absent.—Corona absent, staminodes occasionally present. 1189
— Climber with tendrils.—Inflorescence axillary. Flowers actino-morphic. Ovary stipitate or corona present. Stigma 1, broad.

Passifloraceae

1189. Flowers actinomorphic. Filaments well-developed. Staminodes 5.—Stigmas 3 or 4. 1190
— Flowers more or less zygomorphic. Filaments short. Staminodes ab-sent.—Anthers bent together into a tube, usually appendiculate. Ovary sessile. Stigma 1, rarely 2–5, then stem woody. . . . . **Violaceae**
1190. Herbs. Leaves radical. Staminodes 5, incised. (*Parnassiaceae*).

Saxifragaceae

— Shrubs or trees. Leaves alternate, cauline. Staminodes absent. (*Escalloniaceae*). . . . . **Saxifragaceae**

111

1191. Herbs.—Ovules many. 1192
— Woody plants... 1193
1192. Sepals 4 or 5, valvate, persistent. Petals 4 or 5........ **Saxifragaceae**
— Sepals 2, early deciduous as a cap. Petals 4.—Pacific N. America. (*Eschscholzia*). .... **Papaveraceae**
1193. Stamens usually more than the petals, rarely as many, then epi-petalous, but calyx imbricate, ovules pendulous and fruit dry.
Flacourtiaceae
— Stamens epipetalous.—Thorny shrubs. Calyx valvate. Ovules 4, ascending. Fruit a drupe. .... **Rhamnaceae**
1194. (1152). Ovule 1 per locule................................. 1195
— Ovules 2–more per locule. 1213
1195. Ovule erect or ascending................................. 1196
— Ovule pendulous or descending........................... 1201
1196. Stamens as many as the petals, epipetalous.—Stem woody. Calyx valvate. 1197
— Stamens either as many as the petals and alternipetalous, or more.
1198
1197. Sepals, petals, and stamens 8. Ovary 8-locular.—Socotra. (*Dirach-maceae*). .... **Geraniaceae**
— Sepals, petals and stamens 4 or 5. Ovary 2–4-locular. . **Rhamnaceae**
1198. Stamens inserted outside the disk or on it. Embryo straight. 1199
— Stamens inserted inside the disk. Embryo more or less curved.—Stem woody. Flowers polygamous. Stamens usually 8. Endosperm absent. .... **Sapindaceae**
1199. Flowers actinomorphic or nearly so. Stamens 4 or 5. Endosperm usually present.. 1200
— Flowers zygomorphic. Stamens 10. Endosperm absent.—Flowers bisexual. Style gynobasic. (*Parinari*)............. **Chrysobalanaceae**
1200. Herbs. Petals connate in the middle.—Leaves alternate. Stigmas 2–5. Malaysia to New Zealand. (*Stackhousia, Tripterococcus*).
Stackhousiaceae
— Woody plants. Petals free. .... **Celastraceae**
1201. Stem herbaceous.—Endosperm absent...................... 1202
— Stem woody. 1204
1202. Flowers actinomorphic, 4-merous. Stamens 2–6. Ovary 2-locular. Stigma 1. Fruit a capsule or dry and indehiscent. 1203
— Flowers zygomorphic. Calyx-segments 5. Petals 5, exceptionally 2. Stamens 8. Ovary 3-locular. Stigmas 3. Fruit a schizocarp, rarely a berry. .... **Tropaeolaceae**
1203. Flowers racemose. Calyx imbricate. Fruit a capsule. Embryo curved.
Cruciferae
— Flowers solitary, axillary. Calyx valvate. Fruit dry, indehiscent.

112

Embryo straight................................**Onagraceae**
1204. Corolla imbricate.. 1205
— Corolla valvate or apert................................. 1209
1205. Stamens 4–6. 1206
— Stamens 8 or 10... 1208
1206. Ovary irregularily 20-locular, apex with a hollow tubule, inside with 5 stigmatic lines and a central, free column which simulates a style.—Leaves alternate. Stipules minute. S.E. Asia to N.E. Australia. (*Siphonodontaceae*)..........................**Celastraceae**
— Ovary and style different................................. 1207
1207. Ovule apical, pendulous.—Stipules absent. N. Zealand. (*Brexiaceae*: *Ixerba*)..................................... **Saxifragaceae**
— Ovule basal, erect..............................**Celastraceae**
1208. Distal part of petioles and nodes of inflorescences with annular glands. Stamens 8, 2 free and in 2 bundles of 3 each. Ovary 2-locular.—Guiana (*Barnhartia*). ....**Polygalaceae**
— Petioles and inflorescences without such glands. Stamens 10, connate at least at base. Ovary 5-, sometimes apparently 10-locular. (incl. *Ixonanthaceae*)........................... **Linaceae**
1209. Corolla valvate.. 1210
— Corolla apert.—Bark inside with tough, silky fibres. Petals 4–10, scale-like. Stamens 8–10........................... 1211
1210. Leaves undivided. Fruit a drupe or dry and indehiscent.—Flowers actinomorphic. Stigma 1. Endosperm present. 1212
— Leaves at least partly lobed. Fruit a berry.—Inflorescence umbellate. Stamens 5. Endosperm ruminate. Himalaya to Malaya. (*Gamblea, Hederopsis*). ....**Araliaceae**
1211. Flowers in panicles. Anthers transversally dehiscent. Ovary 3–5-locular. Embryo curved. (also in *Thymelaeaceae*). ... **Gonystylaceae**
— Flowers in umbels or capitules. Anthers longitudinally dehiscent. Ovary 2-locular. Embryo straight...................**Thymelaeaceae**
1212. Leaves opposite. Stipules present.—Moluccas to Fiji. (*Mastixiodendron*). ....**Rubiaceae**
— Leaves alternate. Stipules absent......................**Olacaceae**
1213. (1194). Ovules 2 per locule............................... 1214
— Ovules 3–more per locule. 1244
1214. Ovules erect or ascending. 1215
— Ovules pendulous, or descending, or patent, or one descending and one ascending.. 1222
1215. Stamens 4–more.. 1216
— Stamens 3, less than the petals.—Filaments short or broad. Anthers extrorse. Ovary 2- or 3-locular. Endosperm absent. (also in *Celastraceae*). .... **Hippocrateaceae**

113

1216. Stamens epipetalous. 1217
— Stamens alternipetalous, or more than the petals. 1218
1217. Leaves opposite. Stigmas 2 or 3. Endosperm scanty. . . . **Rhamnaceae**
— Leaves alternate. Stigma 1. Endosperm copious. . . . . **Vitaceae**
1218. Stamens more than petals. 1219
— Stamens as many as the petals.—Leaves undivided. Stipules usually present. Calyx imbricate or apert. Flowers actinomorphic. Stamens inserted on the margin of the disk or close to it. Anthers usually introrse. Endosperm usually present. . . . . **Celastraceae**
1219. Leaves not translucent-glandular-punctate. 1220
— Leaves translucent-glandular-punctate.—Flowers actinomorphic. Disk intrastaminal. . . . . **Rutaceae**
1220. Leaves opposite, simple. 1221
— Leaves alternate, usually compound.—Flowers polygamous. Disk extrastaminal. . . . . **Sapindaceae**
1221. Flowers zygomorphic.—Calyx valvate. . . . . **Lythraceae**
— Flowers actinomorphic.—Leaves usually lobed. Stipules absent. Flowers unisexual or polygamous. Stigmas 2. Fruit a schizocarp. Endosperm absent. . . . . **Aceraceae**
1222. Ovary 4- or 5-locular. 1223
— Ovary 2- or 3-locular. 1231
1223. Flowers actinomorphic. 1224
— Flowers zygomorphic.—Stipules present. Flowers solitary or in umbels. Corolla spurred, spur adnate to the pedicel, inconspicuous. Stamens 2–7(–10). Stigmas 5. Endosperm absent. (*Pelargonium*).
Geraniaceae
1224. Stamens 4 or 5. 1225
— Stamens 8–10. 1228
1225. Leaves not translucent-glandular-punctate. 1226
— Leaves translucent-glandular-punctate.—Leaves alternate, undivided. Stipules absent. Stamens 5, alternipetalous. Endosperm absent. . . . . **Rutaceae**
1226. Leaves opposite or in whorls. Stamens alternipetalous, 4 or 5. . 1227
— Leaves alternate. Stamens epipetalous. Calyx valvate. Petals scale-like. Stamens 5. Stigmas 4 or 5. Endosperm present. . . **Sterculiaceae**
1227. Disk thick. Anthers broad. Stigma lobed. Endosperm copious.
Celastraceae
— Disk thin. Anthers narrow. Stigma undivided. Endosperm scanty.
Saxifragaceae
1228. Leaves alternate, undivided, or compound. 1229
— Leaves opposite, undivided.—Calyx and corolla valvate. Stigma 1. Fruit dehiscent. Endosperm present. (*Cassipourea*, *Macarisia*).
Rhizophoraceae

1229. Leaves compound. Endosperm absent. 1230
— Leaves undivided. Endosperm present.—Calyx and corolla imbricate, persistent. Stigma 1. Fruit a capsule. (incl. *Ixonanthaceae*).

Linaceae

1230. Leaves trifoliolate. Calyx apert. Corolla imbricate. Stigma 4- or 5-lobed. Fruit a berry.—Mauritius, Indomalesia. (*Sandoricum*).

Meliaceae

— Leaves pinnately compound. Calyx and corolla valvate. Stigma 1. Fruit a drupe. . . . . **Burseraceae**

1231. Stigmas 2 or 3. 1232
— Stigma 1. 1235

1232. Calyx imbricate. 1233
— Calyx valvate.—Leaves alternate. Stipules present. Stamens 5, epipetalous. Fruit a capsule. . . . . **Sterculiaceae**

1233. Leaves alternate. Stipules present. 1234
— Leaves opposite. Stipules absent.—Flowers in racemes, unisexual or polygamous. Fruit a winged schizocarp. . . . . **Aceraceae**

1234. Stamens 10. Fruit a capsule, or dry and indehiscent.

Dipterocarpaceae

— Stamens 5. Fruit a drupe.—Flowers in cymes. . . . . . . **Dichapetalaceae**

1235. Stamens 3–10. 1236
— Stamen 1.—Leaves opposite or in whorls, simple, undivided. Flowers zygomorphic. . . . . **Vochysiaceae**

1236. Stamens as many as the petals or more. 1237
— Stamens 3. Petals 5.—Leaves undivided. (also in *Celastraceae*).

Hippocrateaceae

1237. Leaves opposite, undivided. 1238
— Leaves alternate, undivided, or compound. 1239

1238. Petals imbricate, 5. Stamens 5. . . . . **Celastraceae**
— Petals valvate, 4 or 5. Stamens 8–10. (*Macarisia*). . **Rhizophoraceae**

1239. Petals imbricate or contort. 1240
— Petals valvate.—Petals 3 or 4. Stamens 6 or 8. Calyx valvate. Fruit a drupe. . . . . **Burseraceae**

1240. Stipules present. Stamens 3–5 or 10.—Leaves undivided. Flowers bisexual. 1241
— Stipules absent. Stamens 5–9.—Petals 4–7. 1242

1241. Petals 3. Stamens 3 or 4. . . . . **Trigoniaceae**
— Petals 5. Stamens 5 or 10. . . . . **Dipterocarpaceae**

1242. Flowers bisexual. Disk absent.—Leaves pinnately compound. . . 1243
— Flowers polygamous. Disk present.—Endosperm absent. Embryo curved. . . . . **Sapindaceae**

1243. Petals imbricate, clawed. China. . . . . **Bretschneideraceae**
— Petals contort, not clawed. Australia. . . . . **Akaniaceae**

1244. (1213). Anthers with longitudinal or transverse slits. 1245
— Anthers with terminal pores.—Leaves opposite or in whorls, rarely radical. Stamens twice as many as the petals, rarely as many.

Melastomataceae

1245. Calyx valvate. 1246
— Calyx imbricate or apert. 1249
1246. Stigma 1. 1247
— Stigmas 2 – 5.—Endosperm present. Embryo straight. 1248
1247. Leaves undivided. Endosperm absent. Embryo straight. **Lythraceae**
— Leaves palmately compound. Endosperm scanty. Embryo curved.

Bombacaceae

1248. Leaves undivided or lobed, cauline. Stipules present. Stamens 5. Stigmas 3 – 5. . . . . **Sterculiaceae**
— Leaves pinnatifid, subradical. Stipules absent. Stamens 4, or 8 – 10. Stigmas 4, rarely 2.—Herbs. S. America. (*Francoaceae: Francoa*).

Saxifragaceae

1249. Woody plants. 1250
— Herbs.—Petals 4. Stamens 6. Ovary 2-locular. Embryo curved.

Cruciferae

1250. Flowers actinomorphic. 1251
— Flowers zygomorphic. 1255
1251. Stamens as many as the petals and alternipetalous, or less. 1252
— Stamens either 5 and epipetalous, or 7 – 10 and more than the petals.—Leaves usually translucent-glandular-punctate and with a marginal nerve. 1256
1252. Petals valvate, rarely imbricate, then either petals 6–9, or disk cupular and fimbriate, or disk indistinct.—Endosperm usually present. 1253
— Petals imbricate, 4 or 5, or less, if 4 or 5 disk thick and more or less expanded. 1258
1253. Stipules absent. 1254
— Stipules present. (*Brexiaceae*). **Saxifragaceae**
1254. Disk present. (*Escalloniaceae*). **Saxifragaceae**
— Disk absent. (*Philadelphaceae*). **Saxifragaceae**
1255. Leaves opposite, undivided.—Stamens as many as the petals and epipetalous, or more. Ovary 3-locular. 1260
— Leaves alternate, pinnately compound.—Flowers 4-merous. Tropical and S. Africa. . . . . **Melianthaceae**
1256. Stamens 7 – 10, more than the petals. 1257
— Stamens 5, epipetalous.—Leaves alternate. Stigma capitate. S. Africa. (*Heteropyxidaceae*). . . . . **Myrtaceae**
1257. Leaves alternate. Stigmas 3 or 4, subsessile.—Mascarenes. (*Psiloxy-laceae*). . . . . **Myrtaceae**

— Leaves usually opposite. Style usually simple and stigma capitate.

Myrtaceae

1258. Stamens 4 or 5. Anthers introrse, rarely extrorse, then ovary 4- or 5-locular. 1259

— Stamens 3. Anthers extrorse. Ovary 3-locular.—Endosperm absent. (also in *Celastraceae*). .... **Hippocrateaceae**

1259. Leaves present. Disk usually conspicuous. Ovary 1–5-locular.

Celastraceae

— Leaves absent. Disk absent. Ovary 5-locular.—Texas, Mexico. (*Canotia*, also in *Koeberliniaceae*). .... **Canotiaceae**

1260. Petal 1. Stamen 1. Endosperm absent. .... **Vochysiaceae**

— Petals 5. Fertile stamens 6. Endosperm present. .... **Trigoniaceae**

1261. (1151). Ovary 1, undivided or lobed. 1262

— Ovaries 2–more, free or connate at base only. 1303

1262. Ovary 1-locular, sometimes incompletely so. 1263

— Ovary 2–more-locular or nearly so. 1278

1263. Ovule 1.—Shrubs or trees. Leaves usually alternate. Endosperm absent or scanty. 1264

— Ovules 2–more, rarely 1, then herbs or undershrubs, leaves usually opposite, stipules usually present. 1267

1264. Stipules absent. 1265

— Stipules present.—Ovule erect. 1269

1265. Ovule pendulous, adnate to the ovary-wall, or pendulous from a basal funicle. 1266

— Ovule basal, erect.—Tropical Africa. (*Stapfiella*, also in *Flacourtiaceae*). .... **Turneraceae**

1266. Bark usually with black (poisonous!) resin. Staminodes, if present, not petaloid. .... **Anacardiaceae**

— Bark without resin. Staminodes 3–6, petaloid.—Calyx imbricate. Fertile stamens 3–6, epipetalous. Styles 2. New Guinea to New Zealand. .... **Corynocarpaceae**

1267. Trees. Flowers in radiate capitules, connate at base, only the outer with 1–4 petals and then usually strongly zygomorphic. Stamens 7–10. Styles 2. Ovules parietal, 1 or 2. Endosperm present.—S.E. Asia. (*Rhodoleiaceae*). .... **Hamamelidaceae**

— Plants otherwise. Flowers usually actinomorphic or nearly so... 1268

1268. Endosperm present. 1270

— Endosperm absent.—Woody plants. Leaves alternate, undivided. Stipules absent. Flowers in racemes. Anthers extrorse. Styles 4–7. Ovules basal or parietal. Embryo straight. .... **Tamaricaceae**

1269. Calyx valvate. Stamens 4–7, epipetalous. Style 2–4-partite.

Rhamnaceae

— Calyx and corolla imbricate. Stamens 3–9, usually more than the

117

petals, if as many alternipetalous. Styles usually 3, less often simple
with 3 free stigmas. .... **Polygonaceae**

1270. Ovules 2 – more, parietal. 1271

— Ovules basal or central, 1 – more.—Herbs or undershrubs. Leaves
undivided, usually opposite. Stipules present. Embryo more or less
curved. .... **Caryophyllaceae**

1271. Anthers introrse or latrorse, rarely extrorse, then stem woody, and
calyx valvate. 1272

— Anthers extrorse.—Herbs, usually with stalked glands or glandular
hairs. Leaves involute in bud. Calyx imbricate. Stamens as many as
petals, 4 – 8. .... **Droseraceae**

1272. Corolla imbricate or valvate, if contort stamens twice as many as the
petals, 8 – 10. 1273

— Corolla contort. Stamens 5, as many as the petals.—Calyx caducous. .... **Turneraceae**

1273. Styles apical on the ovary, adjacent at base, rarely somewhat distant, then plants woody and stipules present. 1274

— Styles subapical on the ovary, free to base.—Erect or prostrate
herbs. Stipules absent. Stamens 5, as many as the petals. Aril absent. S. America. .... **Malesherbiaceae**

1274. Stem erect. Tendrils absent. Corona usually absent, rarely present,
exceptionally double. Ovary (sub-)sessile. 1275

— Climbers. Tendrils present, plants rarely erect without tendrils.
Corona usually present. Ovary usually stipitate.—Stamens 4 – 6, as
many as the petals, alternipetalous, rarely more. Styles or style-
branches 3, rarely 4 or 5. Aril present. .... **Passifloraceae**

1275. Stem herbaceous. 1276

— Stem woody.—Leaves undivided. 1277

1276. Staminodes absent. .... **Saxifragaceae**

— Staminodes present. (*Parnassiaceae*). .... **Saxifragaceae**

1277. Leaves 3-foliolate, apparently in whorls of 6 leaflets.—Australia.
(*Baueraceae*). .... **Saxifragaceae**

— Leaves undivided, alternate. .... **Flacourtiaceae**

1278. (1262). Ovule 1 per locule. 1279

— Ovules 2 – more per locule. 1290

1279. Ovule erect or ascending. 1280

— Ovule pendulous or descending.—Woody plants. 1283

1280. Stamens as many as the petals, alternipetalous, or more. 1281

— Stamens 4 or 5, epipetalous.—Woody plants. Leaves undivided.
Calyx valvate. Endosperm present. Embryo straight. ... **Rhamnaceae**

1281. Herbs. Flowers bisexual. Embryo straight. 1282

— Woody plants. Flowers polygamous. Embryo curved.—Endosperm
absent. .... **Sapindaceae**

118

1282. Radical leaves undivided, cauline ones pinnatifid. Calyx valvate. Fruit a capsule.—Australia. (*Eremosynaceae*). . . . . **Saxifragaceae**
— Leaves all entire. Calyx imbricate. Fruit a schizocarp or nutlets 2–5. . . . . **Stackhousiaceae**
1283. Filaments free, if connate at base plant glabrous. Leaves alternate and endosperm present. 1284
— Filaments connate at base.—Indument with medifixed hairs. Leaves undivided, opposite. Anthers with longitudinal slits. Fruits indehiscent. Endosperm absent. . . . . **Malpighiaceae**
1284. Leaves compound, alternate. 1285
— Leaves simple, alternate, or opposite. 1286
1285. Stipules present. Stamens as many as the petals. Styles 3. Endosperm absent.—Leaves pinnately compound. . . . . **Araliaceae**
— Stipules absent. Stamens twice as many as the petals. Styles or style-branches 4 or 5. Endosperm present. . . . . **Anacardiaceae**
1286. Ovary irregularly 20-locular, apex with a hollow tubule, inside with 5 stigmatic lines and a central column, which simulates a style.—Leaves alternate. Stipules minute. Flowers axillary, solitary or in cymes. Stamens 5. S.E. Asia to N.E. Australia. (*Siphonodontaceae*). . . . . **Celastraceae**
— Ovary and style different. 1287
1287. Stipules absent, when present leaves opposite and anthers with longitudinal slits. 1288
— Stipules present.—Leaves alternate, rarely opposite, then, as usual, anthers with valves. Flowers in spikes or capitules. Styles 2. Fruit a capsule. Endosperm present. (*Hamamelis, Trichocladus*).
Hamamelidaceae
1288. Leaves alternate. Stipules absent.—Styles 2. 1289
— Leaves opposite. Stipules present.—Flowers solitary or in panicles. Stamens 8–10. Anthers with longitudinal slits. . . . . **Cunoniaceae**
1289. Shrublets. Flowers in capitules. Stamens 5, alternipetalous. Fruit a capsule. Endosperm present. S. Africa. . . . . **Bruniaceae**
— Shrubs or trees. Flowers in panicles. Stamens 3–6, epipetalous. Fruit a drupe. Endosperm absent. S.W. Pacific. . . . **Corynocarpaceae**
1290. (1278). Ovules 2 per locule. 1291
— Ovules 3–more per locule. 1297
1291. Leaves opposite.—Stem woody. 1292
— Leaves alternate. 1293
1292. Stipules absent. Flowers unisexual or polygamous. Fruit a winged schizocarp. Endosperm absent. . . . . **Aceraceae**
— Stipules present. Flowers bisexual. Fruit dehiscent, or dry and indehiscent. Endosperm present. . . . . **Cunoniaceae**
1293. Ovules pendulous.—Leaves undivided. 1294

119

Stamens 5. . . . . **Staphyleaceae**

1305. Flowers bisexual or polygamous. 1306

— Flowers unisexual.—Trees. Leaves pinnately lobed. Stipules antidromous. Flowers in capitules. Stamens as many as the petals. Ovule 1 per carpel. . . . . **Platanaceae**

1306. Herbs. Leaves alternate, compound. Flowers in panicles. Stamens 8 – 10. Carpels adnate at base with the receptacle. Ovules 2 or 3 per carpel. . . . . **Saxifragaceae**

— If not as above, try: . . . . **Rosaceae**

1307. Anthers extrorse or with valves. Carpels many.—Leaves opposite, undivided. Carpels indehiscent. 1308

— Anthers introrse with longitudinal slits. Carpels 3 – 10. 1309

1308. Anthers usually with valves. Ovule 1 per carpel. Endosperm copious. . . . . **Monimiaceae**

— Anthers with longitudinal slits. Ovules 2 per carpal. Endosperm very scanty. China. (*Chimonanthus*). . . . . **Calycanthaceae**

1309. Leaves simple, undivided or lobed. 1310

— Leaves compound.—Stem woody. Leaves alternate. Flowers in racemes or panicles. Carpels 3 – 5, each with 2 collateral ovules.

Connaraceae

1310. Carpels as many as the petals, 4 – 10, rarely 3, then stamens 3. Endosperm absent or very scanty. 1311

— Carpels 2 or 3, less than the petals. Stamens 8 – 10. Endosperm copious.—Herbs. Leaves alternate. . . . . **Saxifragaceae**

1311. Carpels free or connate at base only. . . . . **Crassulaceae**

— Carpels connate to about halfway, 5 – 8.—Carpels circumscissile at the base of the free part. E. Asia, E. N. America. (*Penthoraceae*, sometimes in *Saxifragaceae*). . . . . **Crassulaceae**

1312. (1150). Ovary 1, undivided or lobed. 1313

— Ovaries 2 – more, free or connate at base only. 1362

1313. Ovary 2 – more-locular, or nearly so. 1314

— Ovary 1-locular, or nearly so. 1318

1314. Leaves deeply divided or compound. 1315

— Leaves undivided. 1336

1315. Leaves deeply divided or palmately compound. 1316

— Leaves pinnately compound.—Leaves alternate. Stipules absent. Flowers in a panicle, polygamous. Style 1. Ovary 2- or 3-locular. Ovule 1 per locule. . . . . **Sapindaceae**

1316. Leaves opposite. Stipules absent. Ovary 2-locular. 1317

— Leaves alternate. Stipules present. Ovary 5-locular.—Flowers solitary or in fascicles, bisexual. Filaments connate. Anthers 1-locular. Style 1. Ovules many per locule. . . . . **Bombacaceae**

1317. Flowers solitary, bisexual. Ovules many per locule.—Leaves pal-

mately compound. Styles 2. . . . . **Saxifragaceae**
— Flowers in racemes, unisexual or polygamous. Ovules 2 per locule.—Leaves deeply divided. Styles 2-partite. . . . . **Aceraceae**
1318. Ovules 1 or 2. 1319
— Ovules 3 – more. 1325
1319. Leaves alternate. 1320
— Leaves opposite. 1322
1320. Leaves simple, undivided, or lobed, rarely ternately dissected. . 1321
— Leaves compound.—Petals 3 – 5. Ovule 1. . . . . **Leguminosae**
1321. Leaves translucent-glandular-punctate. Stipules absent. Ovule 1, erect.—Petals 5. Style basal. Calyx with an entire margin. N. Brazil, Guianas. . . . . **Rhabdodendraceae**
— Leaves not translucent-glandular-punctate. Stipules usually present. Ovules 1 or 2.—Style and stigma simple. Shrubs or trees. 1324
1322. Stigma sessile, terminal. 1323
— Style present, 1 – several.—Leaves simple, or compound, or reduced to a widened petiole. 1329
1323. Leaves translucent-glandular-punctate. Ovule pendulous. Endosperm fleshy. . . . . **Monimiaceae**
— Leaves not translucent-glandular-punctate, pusticulate by crystals when dry. Ovule basal. Endosperm absent.—Cotyledons 3 or 4, massive, fleshy. Queensland. (*Idiospermaceae*). . . . . **Calycanthaceae**
1324. Style terminal. Ovules 1 or 2, parietal. (*Prunoideae*). . . . . . **Rosaceae**
— Style gynobasic. Ovules 2, basal. . . . . **Chrysobalanaceae**
1325. Ovules on 1 or more parietal or central placentas. 1326
— Ovules apical on a central column, about 6 in 2 groups, pendulous.—Flowers 5-merous. W. Africa. (also in *Flacourtiaceae, Passifloraceae, Medusagynaceae*). . . . . **Soyauxiaceae**
1326. Placenta 1, parietal.—Leaves alternate. Stipules usually present. Style and stigma simple. 1327
— Placentas 2 or more, or central, rarely apparently parietal, then leaves undivided, nearly always opposite, stipules absent, calyx lobes 6, valvate, and stamens 11. 1329
1327. Leaves compound (leaflets entire) or reduced to a widened petiole. Flowers actinomorphic or zygomorphic.—Ovules 1 or more and serial. . . . . **Leguminosae**
— Leaves simple, entire or serrate, rarely lobed or dissected in 3 lobes. 1328
1328. Ovules 1 or more, collateral. . . . . **Rosaceae**
— Ovules numerous on intrusive parietal placentas. (*Prockieae*, also in *Tiliaceae*). . . . . **Flacourtiaceae**
1329. Ovary distinctly, usually long stipitate. Sepals and petals 4.—Leaves alternate. Style and stigma simple. Embryo curved. . . . . **Capparaceae**

— Ovary sessile or nearly so, rarely shortly stipitate but then petals 5
or more. 1330
1330. Shrubs or woody plants. 1331
— Herbs, sometimes climbing. 1334
1331. Leaves opposite.—Ovary completely divided into locules, at the
base more or less adnate to the receptacle. . . . . **Sonneratiaceae**
— Leaves alternate or absent. 1332
1332. Ovary sessile. Embryo straight.—Endosperm present. **Flacourtiaceae**
— Ovary stipitate. Embryo curved. 1333
1333. Flowers 8-merous, zygomorphic. Stigmas 2 or 3, sessile.—Leafless
shrubs. . . . . **Resedaceae**
— Sepals 2, imbricate. Petals 5. Style simple.—Ovules basal. Endo-
sperm absent. . . . . **Capparaceae**
1334. Style 1. Stigma usually 1, rarely 3–7, then sepals more than 2. 1335
— Stigmas 4–6, (sub-)sessile.—Herbs. Leaves dissected. Sepals 2,
early caducous as a cap. Petals 4, imbricate. Pacific N. America.
(*Eschscholzia*). . . . . **Papaveraceae**
1335. Calyx valvate. Staminodes usually absent. Endosperm absent.—
Ovary not broadly sessile in the receptacle. . . . . **Lythraceae**
— Calyx imbricate or apert. Staminodes present, hollow. Endosperm
present. . . . . **Loasaceae**
1336. (1314). Ovule 1 per locule. 1337
— Ovules 2–more per locule. 1340
1337. Ovules pendulous.—Leaves alternate, rarely opposite. (*Aëtoxylon*).
1338
— Ovules erect, ascending or patent.—Leaves alternate or opposite.
1339
1338. Bark with tough silky fibres inside. Stipules absent. Filaments free.
Ovary 3–8-locular.—Flowers actinomorphic. Petals partite. Fruit a
berry. Endosperm absent. (also in *Thymelaeaceae*). . **Gonystylaceae**
— Bark without such fibres. Filaments connate at base. Ovary 10-
locular. . . . . **Linaceae**
1339. Leaves opposite. Stipules absent. Flowers actinomorphic. **Guttiferae**
— Leaves alternate. Stipules present. Flowers zygomorphic.—Ovary 2-
locular. Style gynobasic. (*Parinari*). . . . . **Chrysobalanaceae**
1340. Ovules 2 per locule. 1341
— Ovules 3 or more per locule. 1351
1341. Flowers unisexual or polygamous.—Leaves opposite. Stipules ab-
sent. . . . . **Aceraceae**
— Flowers bisexual. 1342
1342. Ovules ascending. 1343
— Ovules pendulous, or descending, or patent. 1345
1343. Flowers zygomorphic.—Calyx 6-lobed, valvate. Petals 6, rarely 2 or

4, imbricate. Stamens 11. Style undivided. . . . . **Lythraceae**

— Flowers actinomorphic, 4- or 5-merous. 1344

1344. Leaves opposite. Stipules absent. Endosperm absent. . . . . **Guttiferae**

— Leaves alternate. Stipules present. Endosperm present.—Filaments connate at base. . . . . **Bombacaceae**

1345. Leaves opposite. 1346

— Leaves alternate. 1348

1346. Calyx and corolla valvate. (*Cassipourea*, *Dactylopetalum*).

Rhizophoraceae

— Corolla imbricate. 1347

1347. Calyx valvate. Filaments free. . . . . **Elaeocarpaceae**

— Calyx imbricate. Filaments connate at base. . . . . . , ... **Linaceae**

1348. Calyx valvate.—Corolla imbricate. Filaments free. Ovary 2–7-locular. . . . . **Elaeocarpaceae**

— Calyx imbricate, or contorted, or cupular and entire, or slightly dentate. 1349

1349. Calyx cupular, entire or slightly dentate.—Stipules absent. Petals valvate, 5–8. Ovary 3–6-locular.—Tropical Africa. **Scytopetalaceae**

— Calyx usually partite or divided to some degree.—Stipules present or absent. 1350

1350. Filaments free. Ovary 5-locular.—Calyx usually accrescent, usually imbricate. Petals 5. . . . . **Dipterocarpaceae**

— Filaments connate at base. Ovary 5-locular.—Calyx and corolla imbricate. . . . . **Linaceae**

1351. (1340). Corolla valvate.—Leaves alternate. Calyx apert. Style 1. Endosperm present.—Tropical Africa. . . . . **Scytopetalaceae**

— Corolla imbricate or apert. 1352

1352. Aquatic herbs. Leaves all radical.—Flowers solitary. Petals numerous. Styles numerous or stigmas sessile. Endosperm present.

Nymphaeaceae

— Woody plants or terrestrial herbs. Not all leaves radical. 1353

1353. Leaves alternate. 1354

— Leaves opposite or in whorls. 1357

1354. Sepals valvate.—Leaves oblique. Flowers solitary or in pairs. Petals 5–7. Stamens numerous. Endosperm present. . . . . **Elaeocarpaceae**

— Sepals imbricate or apert. 1355

1355. Stipules present, often early caducous. Petals sepaloid, 3–5. Stigma 1. Endosperm present.—Flowers in panicles or in racemes. Stamens usually numerous, rarely few. (*Flacourtieae*). . . . . **Flacourtiaceae**

— Stipules absent. Petals not sepaloid. Stigmas 2 or more. Endosperm absent. 1356

1356. Stamens 11 or 12. Filaments inserted on a disk on the hypanthium.—Mascarenes. (*Psiloxylaceae*). . . . . **Myrtaceae**

— Stamens usually numerous. Filaments free, not on a disk. **Theaceae**
1357. Stigma 1. 1359
— Stigmas 2 or more. 1358
1358. Flowers usually unisexual. Calyx imbricate. Endosperm absent.—
Leaves often translucent-glandular-dotted or -striped. . . . . **Guttiferae**
— Flowers bisexual. Calyx usually valvate. Endosperm present.—
Shrubs. Flowers in panicles. Petals 5 – 7. Styles 5 – 7 fid. (*Philadelphaceae*).
. . . . **Saxifragaceae**
1359. Calyx valvate. Anthers dehiscing with longitudinal slits. 1360
— Calyx imbricate or apert, or with tardily separating segments, or
with a calyptrate apical part, rarely valvate but then anthers dehiscing with terminal pores.—Leaves opposite or in whorls. 1361
1360. Stamens about twice as many as the petals, if more either herbs or
shrubs, or inflorescences many-flowered. Ovules axillary, in 2-
locular ovaries central on the sept.—Flowers usually more or less
perigynous. . . . . **Lythraceae**
— Stamens many. Ovary 4 – 21-locular, ovules on the septs.—Trees.
Flowers more or less epigynous. (Probably not distinct from
Lythraceae). . . . . **Sonneratiaceae**
1361. Leaves translucent-glandular-punctate, with 1 distinct main nerve,
sometimes with a distinct submarginal vein. Anthers with a small
gland, but without other appendages, nearly always with longi-
tudinal slits. . . . . **Myrtaceae**
— Leaves not translucent-glandular-punctate, usually with 3 – 11 sub-
equal nerves from the base. Anthers with various appendages, nearly
always with 1 or 2 apical pores.—Filaments bent inwards in bud.
Melastomataceae
1362. (1312). Aquatics.—Leaves peltate. Sepals 4. Petals numerous. Ovaries free, sunken in the enlarged receptacle. (*Nelumbonaceae*).
Nymphaeaceae
— Terrestrial plants. 1363
1363. Ovule 1 per carpel. Stipules absent. 1364
— Ovules 2 or more per carpel, if only 1, stipules present. 1367
1364. Carpels 4 or more. 1365
— Carpels 1 or 2.—Embryo with 4 or 5 massive fleshy cotyledons.
Queensland. (*Idiospermaceae*). . . . . **Calycanthaceae**
1365. Flowers in inflorescences. Anthers either with an operculum or with
introrse slits. 1366
— Flowers solitary. Anthers with extrorse slits.—Leaves opposite.
Flowers 4.5 – 7 cm in diameter. China. (*Sinocalycanthus*).
Calycanthaceae
1366. Leaves opposite. Ovule anatropous, basal. Endosperm copious.
Monimiaceae

— Leaves alternate. Ovule atropous, apical. Endosperm scanty. New Caledonia. (also in *Monimiaceae*). .... **Amborellaceae**

1367. Leaves opposite, undivided. Stipules absent. Anthers extrorse. Ovules 2 per carpel.—Shrubs or trees. Flowers solitary. Perianth segments numerous, gradually merging from sepals to petals. Carpels numerous. Endosperm very scanty. .... **Calycanthaceae**

— Leaves compound and then stipules present or absent, or simple, usually alternate. Anthers introrse or latrorse. Ovules 3 or more per carpel, rarely 1 or 2, but then either leaves alternate, or stipules present, or compound without stipules...................... 1368

1368. Stipules present, rarely absent, but then trees or shrubs with flowers in racemes or in panicles. Aril absent................... **Rosaceae**

— Stipules absent. Herbs or undershrubs or climbers, rarely shrubs but then flowers solitary and aril present........................ 1369

1369. Leaves simple, rarely compound but then plant usually succulent.
1370

— Leaves compound. Herbs.—Stipules absent. Carpels 2–5. Endosperm copious. (*Paeonia*, also in *Ranunculaceae*). ..... **Paeoniaceae**

1370. Shrubs. Flowers 5-merous. .... **Crossosomataceae**

— Herbs or undershrubs. Flowers 6–30-merous.—Flowers in cymes or in panicles. Stamens as many or twice as many as petals.
Crassulaceae

1371. (1149). Stamens 1–10. 1372

— Stamens 11 or more. 1519

1372. Stamens as many as petals and epipetalous................... 1373

— Stamens as many as petals and alternipetalous, or more, or less.
1383

1373. Stamen 1. Flowers zygomorphic.—Leaves undivided. Petal 1.
Vochysiaceae

— Stamens 4–9. Flowers actinomorphic or nearly so. 1374

1374. Leaves palmately compound. Filaments nearly completely connate.
Bombacaceae

— Leaves simple and undivided or lobed. Filaments free or connate at base only. 1375

1375. Styles 2–8, free or connate at base, with free stigmas. Ovary 1-locular with 3 or more ovules, rarely with 2 pendulous ovules.   1376

— Style 1, undivided. Stigma undivided or lobed, rarely divided and with several stigmas and then ovary 2–4-locular, rarely 1-locular with 2 erect ovules. 1378

1376. Shrubs or trees. Sepals 4–8.—Anthers extrorse. Endosperm copious. .... **Flacourtiaceae**

— Herbs. Sepals 2 or 5.. 1377

1377. Flowers solitary or in fascicles. Sepals 2. Placenta central. Embryo

curved. (*Portulaca*)............................. **Portulacaceae**
— Flowers in racemes. Sepals 5. Placentas parietal. Embryo straight.

Saxifragaceae

1378. Autotrophic plants. Ovary with 2 – more clearly distinct ovules. 1379
— Green hemi-parasites, usually epiphytic, exceptionally terrestrial. Ovules either fused with each other or even with the ovary-wall.— Corolla valvate. Fruit juicy. .... **Loranthaceae**
1379. Ovary 2 – 5-locular with 1 ovule per locule, rarely 1-locular with 2 – 5 ovules. Calyx and corolla both valvate. 1380
— Ovary 2 – 5-locular with 2 or more ovules per locule, rarely with 1 ovule or ovary 1-locular with 1 or more ovules, but then calyx and corolla imbricate or apert.—Leaves opposite. Stipules absent or very inconspicuous. Stigma 1. Endosperm absent. 1382
1380. Ovules pendulous.—Leaves alternate. Stipules absent. Corolla valvate. Stigma 1. Endosperm copious........................ 1381
— Ovules erect.—Leaves alternate or opposite. Stipules usually present.. **Rhamnaceae**
1381. Flowers unisexual. Stigma 3 – 5-lobed, lobes bifid. (*Octoknemaceae*).

Olacaceae

— Flowers usually bisexual. Stigma not with bifid lobes. .... **Olacaceae**
1382. Leaves more or less glandular-punctate. Corolla imbricate.

Myrtaceae

— Leaves not punctate. Corolla valvate.—Ovary 3 – 5-locular. Ovules 2 or 3 per locule. Fruit a drupe.—Tropical and S. Africa. .. **Oliniaceae**
1383. (1372). Styles 2 or more, free or more or less completely connate, but not up to the stigmas, *or* with several sessile stigmas....... 1467
— Style 1, with 1 stigma or with several stigmas adjacent at base, *or* stigma 1, sessile.. 1384
1384. Ovary 1-locular, sometimes incompletely so. 1385
— Ovary completely or nearly completely 2- or more-locular. 1419
1385. Ovule 1. 1386
— Ovules 2 or more. 1406
1386. Ovules erect. Stamens usually 10, rarely less. 1387
— Ovules pendulous, rarely erect, then stamens 1 – 5. 1388
1387. Climbers or sprawling shrubs with watch-spring hooks. Leaves not translucent-glandular-punctate.—Petals shortly connate or coherent at base. Stigmas 3. Endosperm ruminate. Tropical Africa to W. Malaysia. .... **Ancistrocladaceae**
— Shrubs or trees. Leaves translucent-glandular-punctate.—Leaves usually with a marginal nerve. Calyx imbricate or apert. Corolla imbricate. .... **Myrtaceae**
1388. Anthers with slits or pores. 1389
— Anthers with valves. 1391

1389. Calyx valvate.—Herbs. Corolla imbricate or apert. 1390
— Calyx imbricate or apert. 1393
1390. Flowers 5-merous. .... **Loasaceae**
— Flowers 2-merous. .... **Onagraceae**
1391. Leaves with cystoliths. Tepals in 1 whorl. Stamens less than tepals.
Anthers latrorse, dehiscing with valves opening upwardly. Staminodial
glands less than stamens or absent.—Leaves simple or palmately 5-lobed.
Flowers polygamous. (*Gyrocarpaceae*). .... **Hernandiaceae**
— Leaves without cystoliths. Tepals in 2 whorls. Stamens as many as
the tepals of the outer whorl. Anthers introrse, longitudinally dehiscing with laterally opening valves. Staminodial glands in 1 or 2
whorls. 1392
1392. Leaves either palmately compound or 3- or 5-foliolate. Flowers bisexual. Fruit with 2–4 lateral wings. (*Illigeraceae*). .. **Hernandiaceae**
— Leaves simple. Flowers unisexual. Fruit globose, enclosed in 2 large
bracts or in a fleshy cupule. .... **Hernandiaceae**
1393. Flowers unisexual and monoecious. Anthers extrorse.—Climbing or
prostrate herbs or undershrubs with tendrils. Leaves cordate, an-
gular or lobed. Endosperm absent................. **Cucurbitaceae**
— Flowers bisexual or polygamous, or dioecious. Anthers introrse or
latrorse. 1394
1394. Leaves compound or pinnately partite. 1395
— Leaves undivided or lobed. 1396
1395. Woody plants. Leaves compound. Sepals entire.—Petals valvate.
Tropics... **Araliaceae**
— Herbs. Leaves pinnately partite. Sepals pinnately partite.—Petals
small, broad, with 2 setae. Mediterranean. (*Lagoecia*). **Umbelliferae**
1396. Flowers actinomorphic, not spurred. Fruit not winged. 1397
— Flowers zygomorphic. Calyx spurred. Fruit winged.—Trees. Flowers
bisexual, in panicles. Petal 1. Stamen 1. Endosperm absent. N.
Brazil, Guianas. (*Erisma*)........................ **Vochysiaceae**
1397. Flowers 3–7-merous................................... 1398
— Flowers 2-merous.—Herbs. Flowers in spikes, or in racemes, or in
panicles. Endosperm present. (*Gunneraceae*)........ **Haloragaceae**
1398. Non-resiniferous herbs or shrubs, hispid (often stinging). Flowers
bisexual, in spikes, or in racemes, or in capitules, 4- or 5-merous.
Petals narrow, imbricate or apert. Fruit a capsule or dry and indehiscent. Endosperm absent. America. (*Gronovioideae*). **Loasaceae**
— Plants different. 1399
1399. Resiniferous (very poisonous!) lofty trees, not hispid. Flowers polygamous, in panicles. Petals valvate or imbricate. Fruit a drupe. Endosperm absent. Himalaya to Thailand. (*Drimycarpus*, *Holigarna*).
Anacardiaceae

— Plants different. Endosperm present. 1400
1400. Non-ericoid shrubs or trees. Ovary inferior. Fruit a drupe or a
berry. 1401
— Ericoid shrubs. Ovary hemi-inferior. Fruit dry and indehiscent.—
Flowers in spikes or in capitules, bisexual. S. Africa. (*Berzelia,
Mniothamnea*). . . . . **Bruniaceae**
1401. Corolla imbricate or apert. 1402
— Corolla valvate. 1403
1402. Bracteoles present at the base of the flower. Style undivided. Fruit a
drupe. . . . . **Nyssaceae**
— Bracteoles absent. Styles 3, or style 1, short, and stigmas 3, re-
curved. Fruit a berry.—New Zealand, S. America. (*Griseliniaceae*).
Cornaceae
1403. Leaves alternate, rarely opposite. Flowers bisexual. 1404
— Leaves opposite. Flowers unisexual.—Himalaya to Japan.
(*Aucubaceae*). . . . . **Cornaceae**
1404. Leaves linear-spathulate, tomentose underneath. Pedicels not articulated. Petals with a small scale at base.—W. Pacific to New Zealand. (*Corokia*, also in *Cornaceae, Escalloniaceae*). . . . **Saxifragaceae**
— Leaves otherwise. Pedicels articulated. Petals without a scale at
base. 1405
1405. Flowers in terminal panicles. Petals 4, or 5, or 8, ovate. Filaments
glabrous.—Leaves opposite or alternate. Indomalesia. (*Mas-
tixiaceae*). . . . . **Cornaceae**
— Flowers in axillary cymes. Petals 4–10, narrowly lanceolate to
linear. Filaments usually hairy.—Leaves alternate. . . . . . . **Alangiaceae**
1406. (1385). Flowers bisexual or polygamous. 1407
— Flowers unisexual.—Usually climbing or prostrate herbs with ten-
drils, rarely erect or shrubby. Leaves alternate. Calyx imbricate or
apert. Corolla valvate. Stamens 1–5. Anthers usually extrorse.
Placentas usually several, parietal. Endosperm absent. **Cucurbitaceae**
1407. Stipules present. 1408
— Stipules absent. 1410
1408. Petals imbricate. 1409
— Petals valvate.—Woody plants. Sepals and petals 5–8. Stamens 10–
16. (*Carallia, Ceriops*). . . . . **Rhizophoraceae**
1409. Herbs. Leaves alternate. Sepals 2. Petals 4–6. Stamens 6–10.
Placenta central. (*Portulaca*). **Portulacaceae**
— Woody plants. Leaves alternate or opposite. Sepals, petals, and
stamens 4 or 5. Ovules basal. . . . . **Celastraceae**
1410. Stamens as many as the petals or more. 1411
— Stamens less than the petals, 3.—Leaves alternate. Petals 6, valvate.

Placenta central. Ovules 3, pendulous. Endosperm copious.

Olacaceae

1411. Placentas 2 – several, parietal. 1412
— Placenta 1, parietal, or basal, or central, or apical.—Trees, shrubs, climbers, or rarely undershrubs. Stigma 1, sometimes 2-lobed or 2- or 4-partite.—Calyx valvate, rarely imbricate. 1415
1412. Stamens 4 or 5... 1413
— Stamens 8 – 10. 1414
1413. Trees. Leaves ± opposite, nigrescent, pinninerved. Flowers 4-merous. Petals much longer than the sepals, valvate, linear. (*Escalloniaceae*: *Polyosma*). .... **Saxifragaceae**
— Shrubs. Leaves alternate, not nigrescent, palmatinerved. Petals usually shorter than the sepals, apert, small and scale-like. (*Grossulariaceae*: *Ribes*)................................. **Saxifragaceae**
1414. Herbs, rarely woody, usually hispid and stinging. Inflorescence cymose. Sepals imbricate. Ovary strictly 1-locular.—America, rarely in Africa, Arabia. (*Kissenia*). .... **Loasaceae**
— Shrubs or trees, non-hispid. Inflorescence usually racemose. Sepals valvate or apert, persistent. Ovary several-locular at base. Tropics and subtropics. .... **Styracaceae**
1415. Ovules apical.—Stigma undivided, sometimes 2-lobed or -partite. 1416
— Ovules basal, or central, or parietal.—Stigma either 4-partite, or simple. 1417
1416. Ericoid shrublets. Ovules 4 – 8, pendulous from a central columella. Endosperm copious.—Leaves not translucent-glandular-punctate. Stamens 5. S. Africa. .... **Bruniaceae**
— Non-ericoid woody plants or climbers. Ovules 2 – 12, pendulous from the apex of the locule. Endosperm absent...... **Combretaceae**
1417. Herbs or undershrubs. Stamens 6 – 8. Stigma 4-partite.—Ovules 3 or more. Endosperm present......................... **Onagraceae**
— Shrubs or trees. Stamens 8 or more. Stigma simple........... 1418
1418. Non-ericoid woody plants. Leaves exceptionally translucent-glandular-punctate, broad and usually thick. Staminodes absent. Ovules basal to central. Fruit a berry. (*Memecylaceae*: *Memecylon*, *Mouriri*)..................................... **Melastomataceae**
— Ericoid shrubs. Leaves translucent-glandular-punctate, narrow. Staminodes present. Ovules more or less parietal. Fruit dry, indehiscent.—Australia. (*Chamaelaucieae*). .... **Myrtaceae**
1419. (1384). Ovule 1 per locule................................ 1420
— Ovules 2 – more per locule. 1439
1420. Calyx imbricate or apert. 1421
— Calyx valvate.—Endosperm absent......................... 1430
1421. Corolla imbricate or apert, sometimes adnate to the ovary..... 1422

130

— Corolla valvate or induplicative-valvate. 1432
1422. Stem woody, rarely herbaceous. Flowers usually bisexual or polyga-
mous, rarely dioecious. 1423
— Stem herbaceous, at most woody at base, climbing or prostrate.
Flowers unisexual, rarely bisexual, then stamens less than the petals.
Cucurbitaceae
1423. Leaves simple, undivided. . . . . ·1424
— Leaves pinnately compound.—Flowers in umbels or in racemes.
Stigmas 5 – 8. . . . . **Araliaceae**
1424. Stigma 1. 1425
— Stigmas 2 or 3. 1428
1425. Perianth differentiated into two whorls (calyx and corolla). Stamens
in a whorl. Anthers with longitudinal slits. 1426
— Perianth simple, segments 7 – 10, in a spiral. Stamens 2 – 10, in a spiral.
Anthers with 2 introrse valves.—Flowers in racemes or panicles.
Ovary inferior. Chile. . . . . **Gomortegaceae**
1426. Leaves not translucent-glandular-dotted. Ovary apparently hemi-
inferior, immersed in a disk.—Flowers solitary or in cymes. 1427
— Leaves translucent-glandular-dotted. Ovary inferior.—Flowers soli-
tary or in fascicles. . . . . **Myrtaceae**
1427. Ovary irregularily 20-locular, apex with a hollow tubule, inside with
5 stigmatic lines and a central free column, which simulates a
style.—Leaves alternate. S.E. Asia to N.E. Australia. (*Siphonodon-
taceae*). . . . . **Celastraceae**
— Ovary and style different. . . . . **Celastraceae**
1428. Flowers in racemes or in panicles. Stigmas 3. 1429
— Flowers in a capitate inflorescence. Stigmas 2.—Ericoid shrublets. S.
Africa. . . . . **Bruniaceae**
1429. Bracteoles absent. Flowers unisexual. Anthers dorsifix. New Zea-
land, S. America. . . . . **Cornaceae**
— Bracteoles 2. Flowers bisexual. Anthers basifix. Madagascar.
(*Melanophyllaceae*). . . . . **Cornaceae**
1430. (1420). Trees. Corolla valvate. Ovule basal. S.E. Asia. (*Axinandra*,
also in *Melastomaceae*). . . . . **Crypteroniaceae**
— Herbs, undershrubs, or aquatics. Corolla imbricate or apert. Ovule
axillary or apical. N. Hemisphere. 1431
1431. Herbs or undershrubs, terrestrial or marshy, but not free-floating.
(*Circaea, Gaureae*). . . . . **Onagraceae**
— Floating aquatics.—Leaves rhomboid, basal half entire, upper half
dentate. Petioles swollen. (*Trapaceae*). . . . . **Onagraceae**
1432. (1421). Stem woody, rarely herbaceous, then leaves opposite.
Stamens 3 – 10, as many as the petals or more. Anthers introrse. Endosperm present. 1433

— Stem herbaceous, sometimes woody at base, climbing or prostrate. Leaves alternate. Flowers unisexual, rarely bisexual (*Schizopepon*), then stamens less than the petals. Stamens 1–5. Anthers extrorse. Endosperm absent............................... **Cucurbitaceae**

1433. Stigma simple, clavate or 2- or 3-lobed. 1434
— Stigma undivided, peltate. .... **Saxifragaceae**

1434. Ovary 1–3-locular.—Ovules with a dorsal, or a lateral, or a ventral raphe. 1435
— Ovary 4-locular.—Innovations with stellate hairs. Ovules with a ventral raphe. S. Africa. (*Curtisiaceae*). .... **Cornaceae**

1435. Petals without a scale at the base. 1436
— Petals with a small scale at the base.—Leaves spathulate-linear, tomentose underneath. New Zealand, Australia. (*Corokia*, in *Cornaceae* or *Escalloniaceae*). .... **Saxifragaceae**

1436. Leaves alternate. Ovules with a lateral or a ventral raphe, the micropyle lateral or external............................ 1438
— Leaves usually opposite, rarely alternate but then ovules with a dorsal raphe and internal micropyle. 1437

1437. Stipules absent. Stamens 4 or 5.—Ovules usually with a dorsal raphe and internal micropyle, rarely with a ventral raphe and micropyle external but then flowers in cymose panicles. .... **Cornaceae**
— Stipules present. Stamens 4.—Petals pilose to papillate inside. Moluccas to Fiji. (*Mastixiodendron*)................... **Rubiaceae**

1438. Stipules absent. Flowers in cymes. Stigma 1, undivided or lobed. Ovules with a lateral raphe and micropyle.—Petals very narrow, recurved. Anthers narrow, longer than the filaments. Ovary 2- or 3-locular. .... **Alangiaceae**
— Stipules either adnate to and scarcely distinct from the base of the petiole, or intrapetiolar, or (rarely) absent. Flowers in umbels, or in capitules, or in racemes, or in spikes. Stigmas 2–20. Ovules with an external micropyle..................................... **Araliaceae**

1439. (1419). Ovules 2 per locule, pendulous. 1440
— Ovules 2 per locule, ascending or patent, or more. 1446

1440. Leaves alternate. Stipules absent........................ 1441
— Leaves alternate or opposite. Stipules present. 1443

1441. Leaves simple, undivided.................................. 1442
— Leaves trifoliolate.—Trees or shrubs. Stamens 10. Stigma 4- or 5-lobed. Mauritius, Indo-Malesia (*Sandoricum*). .... **Meliaceae**

1442. Ericoid shrubs or undershrubs. Stamens 5. Stigmas 2 or 3.—S. Africa.. **Bruniaceae**
— Herbs or non-ericoid undershrubs. Stamens 6–8. Stigmas 3 or 4.
                                                          Onagraceae

1443. Leaves alternate. Fruit a drupe, or dry and indehiscent. Endosperm

absent.—Sepals imbricate. Ovary 2- or 3-locular. 1444
— Leaves opposite. Fruit a capsule or a berry. Endosperm present.—
Disk present. Ovary 2–6-locular. 1445
1444. Petals valvate or imbricate. Stamens 5. Nectaries 5, epipetalous.
Fruit a drupe. (*Dichapetalum*). .... **Dichapetalaceae**
— Petals contort. Stamens 10. Nectaries absent. Fruit dry, indehiscent.
(*Vatica*). .... **Dipterocarpaceae**
1445. Sepals and petals valvate. Stamens 8–10. Fruit a capsule or a berry.
Rhizophoraceae
— Sepals and petals imbricate. Stamens 4 or 5. Fruit a capsule. (*Euonymus*). .... **Celastraceae**
1446. Flowers unisexual. 1447
— Flowers bisexual or polygamous. 1449
1447. Leaves opposite. Corolla imbricate or contort. Ovules 4–many per
locule. Fruit a capsule, or dry and indehiscent.—Erect, woody
plants. Ovary 2-locular. 1448
— Leaves alternate. Corolla usually valvate. Ovules 2 or 3 per locule.
Fruit a berry or a nut.—Plants usually climbing with tendrils, or
prostrate. .... **Cucurbitaceae**
1448. Stamens 3–5. Anthers extrorse. Stigma 2-lobed. Ovules 4 or 5 per
locule, on the sept. Fruit dry, indehiscent. E. Africa, Madagascar.
(*Montiniaceae*: *Grevea*). .... **Saxifragaceae**
— Stamens 10. Anthers introrse. Stigma punctiform. Ovules many per
locule, sub-basal-parietal. Fruit a capsule. S.E. Asia, Pacific.
(*Astronia*). .... **Melastomataceae**
1449. Herbs. Corolla valvate. Endosperm present.—Leaves alternate.
Stipules absent. Stamens 5. 1450
— Corolla imbricate or apert, rarely valvate, then stem woody. Plants
either woody, or herbs and endosperm absent. 1451
1450. Leaves strongly asymmetric. Flowers in cincinni.—S.E. Asia to
Malesia. (also in *Campanulaceae*). .... **Pentaphragmataceae**
— Leaves usually symmetric. Inflorescences various, usually capitules,
or panicles, or flowers solitary. .... **Campanulaceae**
1451. Stipules present, sometimes early fugacious.—Stem woody. 1452
— Stipules absent, rarely present, then calyx valvate and corolla imbricate or apert and either plants herbaceous, or stamens 8; sometimes
with an interpetiolary ridge between opposite leaves. 1460
1452. Calyx usually valvate or apert at base only, rarely apert, then leaves
alternate and corolla valvate. 1453
— Calyx imbricate or apert.—Leaves usually opposite. Corolla imbricate. Stamens 4 or 5. Endosperm usually present. ..... **Celastraceae**
1453. Leaves alternate. 1454
— Leaves opposite.—Stamens 4–10, free. 1456

1454. Fertile stamens 5, free. 1455
— Fertile stamens 10, filaments nearly completely connate.—Corolla imbricate. Endosperm sparse to absent. . . . . **Bombacaceae**
1455. Petals valvate. Disk annular. Stamens 5, staminodes absent. Ovary 2-locular. Endosperm present. (*Iteaceae*, also *Escalloniaceae*).

                                                                Saxifragaceae
— Petals imbricate. Disk absent. Stamens 5, staminodes 5. Ovary 5-locular. Endosperm absent.—Mexico. (*Pterostemonaceae*) . . . . . . **Saxifragaceae**
1456. Stamens 8–10. 1457
— Stamens 4–6. 1458
1457. Petals straight in bud, incised or fimbriate, valvate. Ovary 4–10-locular. Ovules axillary, patent, 2 or more per locule. Endosperm present. (*Gynotroches*, *Pellacalyx*). . . . . **Rhizophoraceae**
— Petals in bud curved over the stamens, imbricate. Ovary 4–6-locular. Ovules basal, 1 or 2 per locule. Endosperm absent. S.E. Asia. (*Axinandra*, also in *Melastomataceae*). . . . . . . **Crypteroniaceae**
1458. Ovary 2-locular. Ovules numerous, on the septs. 1459
— Ovary 3–5-locular. Ovules 3 per locule, basal.—Borneo. (*Dactylocladus*, also in *Crypteroniaceae*). . . . . **Melastomataceae**
1459. Midrib of the leaves prominent above. Flowers about 5 mm in diameter. Seeds in 4 rows per ovary. Tropical America. (*Alzatea*, also in *Oliniaceae* or *Crypteroniaceae*). . . . . **Lythraceae**
— Midrib of leaves flat or slightly immersed above. Flowers about 1 mm in diameter. Seeds in 2 vertical rows per ovary. S. Africa. (*Rhynchocalyx*, also in *Crypteroniaceae*). . . . . **Lythraceae**
1460. Anthers with longitudinal slits. 1461
— Anthers with 1 or 2 terminal pores.—Leaves usually opposite or in whorls, usually with several subequal basal nerves. Calyx imbricate, or apert, or calyptrate, rarely valvate. Filaments incurved in bud. Anthers basifix. Stigma 1. . . . . **Melastomataceae**
1461. Leaves not translucent-grandular-punctate, without marginal nerves.

                                                            1462
— Leaves translucent-glandular-punctate, with marginal nerves.—Stem woody. Calyx imbricate or apert. Stigma 1. . . . . **Myrtaceae**
1462. Woody plants without stipules. 1463
— Woody plants with stipules, or plants herbaceous.—Calyx valvate. Anthers dorsifix. . . . . **Onagraceae**
1463. Plants not ericoid. Ovules usually many per locule. Fruit a capsule or a berry. 1464
— Ericoid shrublets. Ovules 4 per locule. Fruit dry and indehiscent.— Stamens 4 or 5. Stigmas 2. Seed 1. S. Africa. . . . . **Bruniaceae**
1464. Disk absent. 1465
— Disk present.—Petals 4 or 5, imbricate or valvate. Stamens 4 or 5.

134

(Escalloniaceae)................................. **Saxifragaceae**

1465. Indument absent or of simple hairs. Inflorescences often with sterile marginal flowers with enlarged, showy sepals. Petals usually valvate.

1466

— Indument usually of stellate hairs. Sterile marginal flowers absent. Petals usually contort. *(Philadelphaceae)*............. **Saxifragaceae**

1466. Inflorescences often with sterile marginal flowers with enlarged showy sepals. Stamens 8–10. Endosperm present. *(Hydrangeaceae)*.

Saxifragaceae

— All flowers fertile and similar. Stamens 5. Endosperm absent.— Especially the petals with red dots and lines. Anthers apically appendiculate. Seeds minute, flat. New Caledonia. *(Platysperma-tion*, not a *Myrtacea* or *Rutacea)*.................. **Saxifragaceae**

1467. (1383). Ovary 1-locular. 1468

— Ovary 2–20-locular. 1487

1468. Ovules 1–4.. 1469

— Ovules 5 or more. 1479

1469. Plants woody, trees or shrubs or less frequently epiphytes or climbers... 1470

— Plants herbaceous or woody at base only, herbs or undershrubs, less frequently climbers or prostrate herbs or aquatic plants........ 1476

1470. Flowers bisexual or polygamous........................... 1471

— Flowers unisexual. 1473

1471. Flowers bisexual. Ovules 2.—Stamens 8–10, styles 2–6.

Flacourtiaceae

— Flowers polygamous. Ovule 1............................. 1472

1472. Non-resiniferous trees. Flowers in panicles or in globose capitules. Stamens 10. Style 2-fid. Fruit samara-like. Endosperm present. China, Tibet. *(Camptotheca*, also in *Cornaceae)*. .... **Nyssaceae**

— Resiniferous (very poisonous!) trees. Flowers in panicles. Stamens 5. Style 1 or 3–5. Fruit a drupe. Endosperm absent. Himalaya to Thailand. *(Drimycarpus, Holigarna)*............... **Anacardiaceae**

1473. Trees or shrubs without tendrils, stem not inflated. 1474

— Woody climbers with tendrils, or shrubs usually with tendrils, or stem inflated and tendrils absent. .... **Cucurbitaceae**

1474. Flowers in racemes or panicles. Stamens 5. Styles 1 or 3–5. Ovule 1. 1475

— Flowers in catkins. Stamens 4. Styles 2. Ovules 2. Warm America. (also in *Cornaceae)*............................ **Garryaceae**

1475. Resiniferous (very poisonous!) lofty trees. Flowers polygamous. Fruit a drupe. Endosperm absent. Himalaya to Thailand. *(Drimy-carpus, Holigarna)*............................ **Anacardiaceae**

— Non-resiniferous trees or shrubs. Flowers dioecious. Fruit a drupe or a berry. Endosperm present. Indo-China, W. Malesia (*Aralidiaceae*, also in *Araliaceae*), or New Zealand, S. America (*Griseliniacea*).

Cornaceae

1476. Ovule 1. 1477

— Ovules 2–5. 1478

1477. Climbing or prostrate herbs, usually with tendrils. Endosperm absent. . . . . **Cucurbitaceae**

— Herbs or undershrubs without tendrils. Endosperm present.—Petals 5. Stamens 5. Styles 2. . . . . **Umbelliferae**

1478. Plants usually climbing with tendrils or prostrate, rarely shrubs or erect herbs. Flowers unisexual, 3- or 5-merous. Fruit a berry or a nut. Endosperm absent.—Styles 3. . . . . **Cucurbitaceae**

— Erect herbs, or undershrubs, or prostrate, or aquatic. Flowers 2- or 4-merous, unisexual or bisexual. Fruit a drupe or a nut. Endosperm present. . . . . **Haloragaceae**

1479. Placenta central.—Herbs. Flowers bisexual. Sepals 2. Petals 4–6. Stamens 6–10. Style 3–8-fid. (*Portulaca*). . . . . **Portulacaceae**

— Placenta parietal or apical. 1480

1480. Flowers bisexual, rarely unisexual but then endosperm copious.— Woody plants. Sepals and petals 4 or 5. 1481

— Flowers unisexual. Endosperm absent. 1486

1481. Stamens 8–10. Anthers extrorse.—Styles 2–6. Placentas several, parietal. . . . . **Flacourtiaceae**

— Stamens 4 or 5. Anthers introrse or latrorse. 1482

1482. Herbs. 1483

— Shrubs or trees. 1484

1483. Leaves opposite. Inflorescence cymose, flowers paired. Ovules apical, pendulous. (*Vahliaceae*). . . . . **Saxifragaceae**

— Leaves radical or alternate, rarely subopposite. Flowers solitary. Ovules parietal.—Perennials or rarely annuals (*Lepuropetalon*) and then leaves succulent. (*Parnassiaceae, Lepuropetalaceae*).

Saxifragaceae

1484. Flowers epiphyllous, from the midrib of a leaf. (*Dulongiaceae*).

Saxifragaceae

— Flowers not epiphyllous. 1485

1485. Shrubs. Leaves lobed. Fruit a berry. (*Grossulariaceae*).**Saxifragaceae**

— Trees, rarely shrubs. Leaves entire or slightly serrate. Fruit a capsule. . . . . **Saxifragaceae**

1486. Plants usually climbing or prostrate, herbaceous, with tendrils, rarely erect or shrubby. Petals 3–6. Stamens 1–5. Styles usually 3. Fruit a berry or dry and indehiscent. . . . . **Cucurbitaceae**

— Trees. Petals in the male flowers 6–8, absent in the female flowers. Stamens or styles 6–8. Fruit a capsule. . . . . **Datiscaceae**

136

cence a raceme or a panicle. Styles 2 or 3. Ovule with a dorsal raphe. 1498
— Petals valvate, when imbricate plants either herbaceous or woody with pinnately compound leaves, rarely simple and then deeply lobed or orbicular and then the pedicels distinctly jointed below the flowers. 1499

1498. Flowers dioecious. Anthers dorsifix.—Indo-China, W. Malesia. (*Aralidiaceae*, also in *Araliaceae*). . . . . **Cornaceae**
— Flowers bisexual. Anthers basifix.—Bracteoles 2, not early fugacious. Madagascar. (*Melanophyllaceae*). . . . . **Cornaceae**

1499. Plants usually herbaceous.—Leaves alternate, usually pinnately compound, rarely entire or palmately nerved to -compound. Stipules absent, leaf-sheaths often well-developed. Flowers 5-merous, usually in compound umbels, rarely in capitules or racemes. Petals usually with an incurved apex. Style-cushions usually 2 or bilobed. Styles 2, free. Ovule apical, pendulous, epitropous. Fruit a schizocarp, rarely a nut. 1500
— Plants woody, rarely herbaceous, then either stipular sheath distinct (*Stilbocarpa, Araliaceae*), or leaves in whorls of 3 or 4 (*Panax, Araliaceae*). 1501

1500. Fruits with a membranous endocarp, mesocarp with parallel resinous canals. . . . . **Umbelliferae**
— Fruits with a woody endocarp, mesocarp without resinous canals, or only in the primary ribs.—Central axis of the fruit not becoming free. (*Hydrocotylaceae*). . . . . **Umbelliferae**

1501. Stipules or stipular structures absent, but an inter-petiolar ridge sometimes present.—Flowers usually in a raceme or in a panicle.
1502
— Stipules or an inter-petiolar stipular sheath present.—Flowers usually in racemose umbels, or in capitules, or in spikes. Style-cushions usually undivided. Fruit a drupe, rarely a berry, or a nut, or a schizocarp, endocarp usually indurated. Ovule epitropous. . . . . **Araliaceae**

1502. Flowers unisexual. 1503
— Flowers bisexual. 1504

1503. Flowers 5-merous. Ovary 4-locular. Himalaya, China. (*Toricelliaceae*). . . . . **Cornaceae**
— Flowers 4-merous. Ovary 2-locular. Madagascar. (*Kaliphora*).
Cornaceae

1504. Ovary 4-locular. Ovules with a ventral raphe.—Young parts with stellate hairs. S. Africa. (*Curtisiaceae*). . . . . **Cornaceae**
— Ovary 1–3(–5)-locular. Ovules with a dorsal raphe. . . . . . . **Cornaceae**

1505. (1487). Ovules 2 per locule. Woody plants. 1506
— Ovules 2 or more per locule, rarely 2, then plants herbaceous and

138

climbing with tendrils and anthers extrorse. 1511
1506. Ovules ascending.—Leaves alternate. Stamens 10. . . . . **Rosaceae**
— Ovules pendulous. 1507
1507. Stamens 6–10. 1508
— Stamens 5.—Leaves undivided. Anthers introrse. 1510
1508. Leaves opposite.—Stamens 8–10. . . . . **Cunoniaceae**
— Leaves alternate. 1509
1509. Stamens 6–8. W. Malesia. (*Anisophylleaceae*: *Combretocarpus*).
Rhizophoraceae
— Stamens 10. S. China, Indochina. (*Mytilaria*). . . . . . **Hamamelidaceae**
1510. Shrubs or trees. Stipules present. Flowers bisexual, rarely unisexual,
in cymes. Fruit a drupe. Endosperm absent. . . . . **Dichapetalaceae**
— Ericoid shrublets. Stipules absent. Flowers bisexual, in capitules, or
in spikes, or in racemes. Fruit a capsule or a nut. Endosperm co-
pious.—S. Africa. . . . . **Bruniaceae**
1511. Flowers unisexual.—Endosperm absent. 1512
— Flowers bisexual or polygamous. 1513
1512. Plants usually climbing with tendrils or prostrate. Corolla usually
valvate. Ovules 2 or 3 per locule. Fruit a berry or a nut.
Cucurbitaceae
— Erect shrubs. Corolla imbricate. Ovules 10–12 per locule. Fruit a
capsule.—S. Africa. (*Montiniaceae*: *Montinia*). . . . . **Saxifragaceae**
1513. Trees. Stipules present.—Flowers bisexual and in spikes, or polygamous and in capitules. Calyx undivided. Stamens 5 or 10. Anthers
with valves.—Queensland. (*Neostrearia*). . . . . **Hamamelidaceae**
— Erect herbs or woody plants, rarely prostrate or climbing. Stipules
absent, if present scale-like and plant grass-like. 1514
1514. Stamens 2 or 3. Fruit dry, indehiscent.—Herbs, forming tussocks.
Sepals 5–7. Petals 5–10. Disk extra-staminal. Temperate and
(sub)antarctic S. Hemisphere. (*Donatia*, also in *Saxifragaceae* or in
Donatiaceae). . . . . **Stylidiaceae**
— Stamens 4–10. Fruit a capsule. Anthers introrse or latrorse. 1515
1515. Herbs. . . . . **Saxifragaceae**
— Shrubs, or trees, or woody climbers. 1516
1516. Flowers not epiphyllous. 1517
— Flowers fasciculate on the midrib of a leaf.—S. America. (*Dulon-
giaceae*). . . . . **Saxifragaceae**
1517. Disk absent. 1518
— Disk present.—Flowers 5-merous. Petals valvate. Styles 2. (*Itea*, in
Iteaceae or *Escalloniaceae*). . . . . **Saxifragaceae**
1518. Indument absent, or hairs simple. Inflorescence sometimes with
sterile marginal flowers with enlarged sepals. (*Hydrangeaceae*:
Hydrangea). . . . . **Saxifragaceae**

— Indument usually of stellate hairs. Sterile marginal flowers absent. (*Philadelphaceae*). .... **Saxifragaceae**

1519. (1371). Style 1, undivided, stigma 1 or 2 – more, adjacent at base, or stigma 1, sessile. 1520

— Styles 2 – more, free, or connate, but stigmas free, or stigmas 2 – more, sessile. 1554

1520. Stigma 1, undivided or lobed. 1521

— Stigmas 2 – more, or 1, then deeply divided. 1544

1521. Ovary 1-locular. 1522

— Ovary 2 – more-locular. 1531

1522. Plants non-parasitic, autotrophous, green. 1523

— Parasite, consisting of a rhizome and a single flower of which the bracts resemble a calyx. .... **Rafflesiaceae**

1523. Ovule 1. 1524

— Ovules 2 – more. 1525

1524. Flowers bisexual, in cymes. Petals valvate or more or less contort, linear-lanceolate. Stigma lobed...................... **Alangiaceae**

— Flowers polygamous-dioecious. Petals imbricate, ovate to oblong. Style bifid. .... **Nyssaceae**

1525. Ovules 2 or 3. 1526

— Ovules many.—Calyx imbricate or apert. 1529

1526. Stipules absent. 1527

— Stipules present. (*Rhizophoreae*).................. **Rhizophoraceae**

1527. Ovules basal, or parietal, or central. 1528

— Ovules apical. .... **Combretaceae**

1528. Leaves translucent-glandular-punctate. Ovules basal, or parietal, or central. .... **Myrtaceae**

— Leaves not translucent-glandular-punctate. Ovules basal. ... **Rosaceae**

1529. Plants usually herbaceous, rarely shrublets or trees.—Leaves entire, or dentate, or lobed, or pinnatifid. 1530

— Woody plants. Leaves alternate.—Leaves undivided. Inflorescence cymose. Stamens 12 – 16, connate at base. Ovary inferior, locular at base. Ovules partly ascending, partly descending. Stigma indistinct. **Styracaceae**

1530. Flowers umbellate, outer flowers sterile with enlarged sepals. Stamens many.—Perennial herbs or undershrubs. Leaves with a bifidly lobed apex. Ovary hemi-inferior, incompletely 5-locular. Ovules patent. Stigma lobed. China, Japan. (*Deinanthe, Hydrangeaceae*). .... **Saxifragaceae**

— Inflorescence without an outer whorl of sterile flowers. Stamens up to 20, usually in epipetalous groups alternating with usually scale-like staminodes.—Herbs, usually hispid and stinging, rarely woody (*Mentzelia*). Stipules rarely present. Placentas 3 – 5, parietal. Ovules many. Mainly American, rarely S.W. Africa or Arabia (*Kissenia*).

Loasaceae

1531. Corolla valvate... 1532
— Corolla imbricate or apert................................ 1535
1532. Stipules absent. Calyx apert. 1533
— Stipules present. Calyx valvate.—Shrubs or trees. Endosperm present. (*Rhizophoreae*). .... **Rhizophoraceae**
1533. Leaves alternate. 1534
— Leaves opposite.—Ovules many. China, S.E. N. America. (*Hydrangeaceae: Decumaria*). .... **Saxifragaceae**
1534. Ovule 1 per locule................................. **Alangiaceae**
— Ovules 2–6(–many) per locule.................. **Scytopetalaceae**
1535. Stipules present.—Flowers solitary. 1536
— Stipules absent....................................... 1539
1536. Herbs... 1537
— Trees.—Stellately hairy. Sepals and petals 4 or 5. Stamens many. C. America (*Dicraspidia*) or Peru (*Neotessmannia*). .... **Tiliaceae**
1537. Herbs. Flowers solitary. 1538
— Trees. Flowers in a thyrse.—Sepals and petals 4. Stamens 16–25. Ovary 2-locular. E. Brazil.................. **Dialypetalanthaceae**
1538. Sepals 4. Petals, stamens, and locules of the ovary many.

Nymphaeaceae

— Sepals, petals, and locules of the ovary 6. Stamens 12. (*Ludwigia*).

Onagraceae

1539. Leaves not translucent-glandular-punctate................... 1540
— Leaves translucent-glandular-punctate.—Shrubs or trees. Leaves usually with a marginal nerve. Calyx imbricate, or apert, or closed and calyptrate. Endosperm absent..................... **Myrtaceae**
1540. Anthers with longitudinal slits, connective inappendiculate.—Woody plants. Stamens numerous................................. 1541
— Anthers usually with terminal pores, connective usually appendiculate at base.—Leaves opposite or in whorls, usually 3–more-plinerved. Petals usually imbricate or contort, rarely valvate. Endosperm absent. .... **Melastomataceae**
1541. Bracteoles present. Petals imbricate. Fruit a berry or a capsule with a lid. Endosperm absent. 1542
— Bracteoles absent. Petals contort. Fruit a longitudinally dehiscent capsule. Endosperm present.—Leaves opposite. Filaments free. Anthers latrorse. (*Philadelphaceae: Philadelphus*)..... **Saxifragaceae**
1542. Leaves usually opposite. Flowers solitary or in fascicles. Stamens free. Anthers introrse, dorsifix. Placentas initially basal, later at least parietal and superimposed. .... **Punicaceae**
— Leaves alternate. Flowers in racemes. Stamens more or less dis-

tinctly connate at base. Anthers latrorse, basifix. Ovules axillary, or
apical, or basal. 1543
1543. Flowers usually zygomorphic. Sepals valvate. Petals 4, or 6, or 8.
Filaments connate at base into an often very unilaterally developed
androphore. Fruit with a lid, either berry-like, or dry. Tropical
America. . . . . **Lecythidaceae**
— Flowers actinomorphic. Sepals ± imbricate, or calyx tearing irregularily at anthesis. Petals 4. Filaments connate at base to nearly free,
not on a unilateral androphore. Fruit a berry without a lid, or 4-
winged, dry and indehiscent (*Combretodendron*). Old World trop-
ics. (*Barringtoniaceae*). . . . . **Lecythidaceae**
1544. (1520). Ovary 1-locular. 1545
— Ovary 2–20-locular. 1546
1545. Sepals 2. Petals 4–6. Ovules basal or central. Fruit a capsule.—
Herbs. Leaves well-developed. Stipules present. (*Portulaca*).
Portulacaceae
— Sepals and petals 4–more, usually many. Ovules parietal. Fruit a
berry.—Succulents, often spiny. Leaves usually scale-like or absent.
Cactaceae
1546. Ovules 2–more per locule. 1547
— Ovule 1 per locule.—Fruit a berry or a drupe. 1553
1547. Ovules many per locule. 1548
— Ovules 2–4 per locule.—Trees. Stipules present. Anthers with
pores. Fruit dry. Endosperm absent. . . . . **Dipterocarpaceae**
1548. Petals many. 1549
— Petals usually 6 or less. 1550
1549. Terrestrial herbs or undershrubs. Fruit a capsule. Embryo curved.
(*Mesembryanthemum*). . . . . **Aizoaceae**
— Aquatic herbs. Fruit a berry. Embryo straight.—Ovules on the
septs. . . . . **Nymphaeaceae**
1550. Leaves alternate. 1551
— Leaves opposite.—Shrubs. Stipules absent. Corolla imbricate or
contort. Stamens 3–6. Anthers with longitudinal slits. Stigmas 3–7.
Fruit a capsule. Endosperm present. (*Philadelphaceae*).
Saxifragaceae
1551. Herbs or undershrubs. 1552
— Trees.—Stipules absent. Corolla valvate. Anthers with pores. Fruit
a drupe. Endosperm present. **Scytopetalaceae**
1552. Stipules absent. Fruit follicular.—Petals 3, minute. Stamens 12. W.
China. (*Saruma*). . . . . **Aristolochiaceae**
— Stipules present. Fruit capsular.—Petals 3–6, imbricate. Stamens
with longitudinal slits. Stigmas 6. Endosperm absent. . . . **Onagraceae**
1553. Leaves opposite, simple. Stipules absent. Perianth simple, segments

142

7–10, imbricate, in a spiral. Stamens 11, in a spiral. Anthers with 2 introrse valves. Stigma 2- or 3-partite. Ovary 2- or 3-locular.—Chile.

Gomortegaceae

— Leaves alternate, incised to compound. Stipules present, sometimes intra-petiolar or adnate to the petiole and inconspicuous. Perianth differentiated into a calyx and corolla. Petals valvate, in a whorl. Stamens in a whorl. Anthers with longitudinal slits. Stigmas 5–more. Ovary 5–more-locular. . . . . **Araliaceae**

1554. (1519). Ovary 1-locular. 1555

— Ovary 2–more-locular. 1560

1555. Flowers bisexual. 1556

— Flowers unisexual.—Herbs. Stipules present. Sepals and petals (4 or) 5. Placenta parietal. Endosperm absent. Embryo straight. Hawaii. (*Hillebrandia*). . . . . **Begoniaceae**

1556. Placentas parietal, sometimes protruding into the locule. 1557

— Placenta central.—Herbs. Stipules present. Sepals 2. Petals 4–6. Endosperm present. Embryo curved. (*Portulaca*). . . . . **Portulacaceae**

1557. Petals imbricate. 1558

— Petals valvate.—Undershrubs. Endosperm scanty. (*Philadelphaceae*). . . . . **Saxifragaceae**

1558. Woody plants without stinging hairs. Endosperm copious. 1559

— Usually herbs, rarely shrublets (*Mentzelia*) or woody climbers (*Fuertesia*), usually hispid and stinging. Endosperm scanty to absent.—Leaves usually divided. Anthers introrse. Mainly American, rarely from S.W. Africa or Arabia (*Kissenia*). . . . . **Loasaceae**

1559. Parietal placentas slightly protruding into the locule.—Leaves undivided. Anthers extrorse. . . . . **Flacourtiaceae**

— Parietal placentas protruding far into the locule.—Undershrubs.

Saxifragaceae

1560. Ovule 1 per locule. 1561

— Ovules 2–more per locule. 1563

1561. Ovule pendulous.—Endosperm present. 1562

— Ovule ascending.—Corolla imbricate. Endosperm absent. . **Rosaceae**

1562. Trees, or shrubs, or undershrubs. Leaves alternate, usually compound or divided. Fruit a berry or a drupe.—Corolla valvate. Styles 5–more. . . . . **Araliaceae**

— Undershrubs. Leaves opposite, divided. Fruit a capsule.—Stamens 12. W. N. America. (*Philadelphaceae*: *Whipplea*). . . . . **Saxifragaceae**

1563. Flowers bisexual, rarely polygamous. 1564

— Flowers unisexual.—Herbs. Stipules present. Ovules many. Fruit a capsule. Endosperm absent. . . . . **Begoniaceae**

1564. Trees, shrubs, rarely undershrubs. Petals 2–10. 1565

— Herbs or undershrubs. Petals numerous.—Ovules many. Endosperm present. 1571

1565. Leaves opposite or alternate. Stipules absent. Ovules 1 or 2 or many. 1566
— Leaves alternate. Stipules present. Ovules usually 2 per locule, rarely 1 ascending, then corolla imbricate and endosperm absent (*Rosaceae*).—Petals 5. 1569

1566. Leaves usually opposite. Stamens not in epipetalous groups. Ovules usually many. Fruit a capsule. Endosperm copious. 1567
— Leaves alternate. Stamens in epipetalous groups. Ovules 1 or 2 per locule. Fruit dry, indehiscent. Endosperm absent.—Shrubs or under-shrubs. Leaves lobed. Stamens many. Styles 2- or 3-fid. America. (*Mentzelia*). . . . . **Loasaceae**

1567. Leaves simple, sometimes deeply lobed. Disk absent. 1568
— Leaves 3-foliolate, apparently in whorls of 6 leaflets. Disk present.
— Leaves opposite, apparently 6 in whorl. Australia. (*Baueraceae*).
Saxifragaceae

1568. Indument absent or of simple hairs. Inflorescence sometimes with sterile marginal flowers with enlarged sepals.—Rhizomatous herbs, or shrubs, or trees. (*Hydrangeaceae*). . . . . **Saxifragaceae**
— Indument generally of stellate hairs. Sterile marginal flowers never present.—Shrubs, sometimes prostrate. (*Philadelphaceae*).
Saxifragaceae

1569. Petals apert, ligulate, fleshy. Fruit a capsule.—China, Indo-China. (*Mytilaria*). . . . . **Hamamelidaceae**
— Petals imbricate or apert. Fruit indehiscent. 1570

1570. Corolla contort. Connective usually with distinct apical appendages. Anthers basifix. Style-branches either shorter than the connate part of the style or stigmas 3, subsessile. Fruit dry, indehiscent, with 2 or 3 enlarged sepals.—S.E. Asia. (*Anisoptera*). . . . . . . **Dipterocarpaceae**
— Corolla usually imbricate, rarely contort. Connective without ap-pendages. Anthers dorsoversatile. Style-branches usually longer than the connate part of the style. Fruit a berry or a drupe. (*Pomoideae*). . . . . **Rosaceae**

1571. Aquatics. Ovules parietal. Fruit composed of berries. Embryo straight.—Leaves all radical. (incl. *Euryaliaceae: Euryale*).
Nymphaeaceae
— Terrestrials. Ovules basal, or parietal, or axillary. Fruit a capsule. Embryo curved. (*Mesembryanthemum, Orygia*). . . . . **Aizoaceae**

SYMPETALAE

1572. (159). Ovary superior or nearly so. 1573

144

— Ovary inferior or hemi-inferior. 2003

1573. Corolla actinomorphic (especially when contort lobes somewhat un-
equal-sided, but equal to each other). 1574

— Corolla more or less zygomorphic. (See glossary). 1896

1574. Stamens free from the corolla, sometimes adherent, but then bases
of the filaments free. 1575

— Stamens adnate to the corolla. 1654

1575. Herbs. Corolla-lobes and stamens many. Styles 5. Ovary 5-locular.
Ovules many per locule. (*Orygia*). .... **Aizoaceae**

— Plants otherwise.. 1576

1576. Outer petals connate, inner petals smaller and free from each other.
(*Exospermum, Bubbia, Zygogynum*). .... **Winteraceae**

— All petals connate and in one whorl. 1577

1577. Fertile stamens as many as the corolla-segments or less. 1578

— Fertile stamens more than the corolla-segments. 1610

1578. Ovary 1, 1-locular. 1579

— Ovary 1, 2–more-locular, or ovaries 2–more, free. 1593

1579. Ovule 1. 1580

— Ovules 2–more. 1585

1580. Ovule basal. 1581

— Ovule apical or parietal. 1584

1581. Stigma 1. 1582

— Stigmas 5.—Flowers bisexual, 5-merous. Disk absent. Stamens
epipetalous. (*Plumbagineae*). .... **Plumbaginaceae**

1582. Filaments free. 1583

— Filaments connate at base.—Leaves opposite. Fruit dry, indehiscent.
Nyctaginaceae

1583. Leaves alternate. Flowers bisexual or polygamous, 5-merous. Disk
present.—Stamens alternipetalous. Fruit a drupe. Plants resiniferous
(often poisonous!). .... **Anacardiaceae**

— Leaves radical. Flowers unisexual, 4-merous.—Herbs, non-
resiniferous. Disk absent. (*Littorella*). .... **Plantaginaceae**

1584. Leaves in whorls. Flowers bisexual. Anthers with longitudinal slits
or apical pores. Disk more or less distinct. Fruit a capsule or a nut.
Embryo straight.—Flowers 4-merous. Stigma 1. .... **Ericaceae**

— Leaves alternate. Flowers unisexual. Anthers with transverse slits.
Disk absent. Fruit a drupe. Embryo curved. ...... **Menispermaceae**

1585. Ovules either 2–more on 1 parietal placenta or 4–more on a central or basal placenta. 1586

— Ovules many on 2–5 parietal placentas. 1588

1586. Placenta central or basal.—Woody plants. Leaves undivided. ... 1587

— Placenta parietal. 1591

1587. Leaves opposite. Flowers 5-merous, in fascicles or in cymes.

Anthers introrse. Style 1. Stigmas 2–4.—Ovules 4, central.

Celastraceae

— Leaves alternate. Flowers 4- or 5-merous, in racemes. Anthers extrorse. Styles 2–5, free or connate at base. (*Tamariceae*).

Tamaricaceae

1588. Leaves alternate. 1589

— Leaves opposite.—Herbs or undershrubs. Petals often connate in the middle. Flowers polygamous. Stigmas 3 or 4. .... **Frankeniaceae**

1589. Stipules absent... 1590

— Stipules present.—Flowers bisexual or polygamous, 5-merous. Filaments connate. .... **Violaceae**

1590. Trees or shrubs, often climbing. Leaves leathery, undivided. Flowers bisexual or polygamous. Stigma 1.......... **Pittosporaceae**

— Herbaceous liana. Leaves membranous, lobed. Flowers monoecious. Stigmas 3–10, as many as the placentas. S. Africa. (*Ceratiosicyos*).

Achariaceae

1591. Leaves compound or reduced to a broadened petiole......... 1592

— Leaves simple, undivided or lobed.—Stem woody. Stipules absent. Flowers usually unisexual. Sepals connate. Ovules 2, pendulous. Fruit a drupe, or dry and indehiscent................. **Icacinaceae**

1592. Stipules usually present. Corolla-lobes valvate. Ovules ascending, anatropous.—Stigma 1......................... **Leguminosae**

— Stipules absent. Corolla-lobes imbricate. Ovules 2, ascending, atropous.—Stem woody........................... **Connaraceae**

1593. (1578). Ovary 1.. 1594

— Ovaries and styles 2–more, free. 1599

1594. Style 1. Stigmas 1 or 2–5, adjacent at base. 1595

— Styles 3–8, free or connate at base but not up to the free stigmas.

1596

1595. Leaves simple, undivided or lobed. Filaments free or connate at base only....................................... 1600

— Leaves usually pinnately compound. Filaments nearly completely connate. .... **Meliaceae**

1596. Stem woody. 1597

— Stem herbaceous, if woody leaves simple and staminodes absent.— Flowers bisexual. (*Linum*). .... **Linaceae**

1597. Flowers unisexual or polygamous. Leaves simple............. 1598

— Leaves pinnately compound or unifoliolate. Flowers bisexual.—Fertile stamens alternating with alternipetalous staminodes. (*Averrhoaceae*)...................................... **Oxalidaceae**

1598. Flowers unisexual or polygamous. Stamens alternipetalous.

Ebenaceae

— Flowers unisexual. Stamens epipetalous............. **Euphorbiaceae**

1599. Leaves alternate, compound.—Stem woody. Flowers 5-merous. Disk absent. Ovules 2, collateral, atropous................**Connaraceae**
— Leaves opposite, undivided. Ovules usually many......**Crassulaceae**
1600. Stigmas 2–5...1601
— Stigma 1, undivided or lobed............................1603
1601. Ovules 2–more per locule............................1602
— Ovule 1 per locule.—Flowers 5-merous. Herbs. Petals usually free at base, connate above. Disk present. Malesia to New Zealand.

Stackhousiaceae

1602. Woody plants. Leaves opposite. Corolla imbricate. Ovules 2 per locule..**Celastraceae**
— Herbs. Leaves alternate, rarely opposite. Corolla valvate. Ovules many per locule..............................**Campanulaceae**
1603. Disk present, rarely indistinct.........................1604
— Disk absent.—Flowers 5-merous. Anthers with 2 longitudinal slits or with terminal pores. Ovary 2(–5)-locular..................1606
1604. Anthers with 1 or 2 terminal pores or 2 longitudinal slits.......1605
— Anthers with 1 longitudinal slit.—Stem woody. Leaves alternate. Flowers 5-merous. Sepals free. Anthers inappendiculate. Ovules numerous. ....·.......**Epacridaceae**
1605. Corolla imbricate.....................................1609
— Corolla valvate.—Climbing shrubs. Leaves opposite or in whorls. Calyx valvate. Anthers extrorse. Ovary 5–7-locular. Fruit a berry. Mauritius. (*Roussea*, also in *Brexiaceae* or *Escalloniaceae*).

Saxifragaceae

1606. Bark inside without tough, silky fibres. Ovules 2–many per locule.

1607

— Bark inside with tough, silky fibres. Ovule 1 per locule.—Woody plants. Flowers in umbels. Corolla annular. (incl. *Aquilariaceae*: *Gyrinops*, *Octolepis*, the latter sometimes in *Flacourtiaceae*).

Thymelaeaceae

1607. Woody, autotrophous plants. Petals imbricate...............1608
— Insectivorous herbs.—Leaves circinnate when young, glandular. Petals contort. Ovules many per locule. Australia......**Byblidaceae**
1608. Ovules 2 per locule, collateral. Endosperm scanty. Sumatra to S. China.....................................**Pentaphylacaceae**
— Ovules many per locule. Endosperm copious........**Pittosporaceae**
1609. Anthers usually with terminal pores, rarely with longitudinal slits, then flowers usually 4-merous, if 5-merous leaves opposite.— Anthers often appendiculate.........................**Ericaceae**
— Anthers with 2 longitudinal slits.—Leaves alternate. Flowers 5-merous. Anthers inappendiculate. Tasmania, Fuegia, Patagonia. (*Prionotaceae*).................................**Epacridaceae**

147

1610. (1577). Stamens twice as many as the corolla-lobes or less. 1611
— Stamens more than twice as many as the corolla-lobes......... 1612
1611. Stamens 4–10. 1613
— Stamens many.—Herbs. Petals many. Styles 5. Ovary 5-locular. Ovules many per locule. (*Corbichonia*). .... **Aizoaceae**
1612. Stamens 12–more... 1634
— Stamens 9.—Calyx- and corolla-lobes 3. Anthers with valves. Ovary with 1 ovule. .... **Lauraceae**
1613. Style 1 *per flower*, stigma 1, or 2–more, then adjacent at base. Ovary 1, if more, more or less connate at least at the apex..... 1614
— Styles 2–more *per flower*, free or connate at base but not up to the stigmas, sometimes ovaries free or connate at base only. 1629
1614. Ovary 1-locular, or incompletely so. 1615
— Ovary 2–more-locular, or nearly so, or ovaries 2–more, more or less connate at least at the apex........................... 1622
1615. Ovule 1. 1616
— Ovules 2–more... 1619
1616. Ovule apical or parietal. 1617
— Ovule basal.—Leaves usually opposite. Filaments connate at base. Endosperm present.............................. **Nyctaginaceae**
1617. Flowers bisexual. Anthers dehiscing longitudinally or apically. Endosperm present or not.—Leaves alternate or in whorls, rarely opposite.. 1618
— Flowers unisexual. Anthers with transverse slits. Endosperm present.—Leaves alternate. Filaments completely connate.
Menispermaceae
1618. Bark inside without tough, silky fibres. Leaves in whorls. Flowers 4-merous. Stamens 6–8. Anthers longitudinally or apically dehiscent. Endosperm present........................... **Ericaceae**
— Bark inside with tough, silky fibres. Leaves alternate, rarely opposite. Flowers 5-merous. Stamens 10. Anthers longitudinally dehiscent. Endosperm absent......................... **Thymelaeaceae**
1619. Leaves alternate. 1620
— Leaves opposite.—Sepals valvate. Petals usually free at base, connate above. Stamens usually 6. Anthers extrorse. Stigmas 2–6. Ovules several–many, on several parietal placentas. . **Frankeniaceae**
1620. Stipules absent. Calyx and corolla imbricate.—Stamens 10. 1621
— Stipules usually present. Corolla- and usually calyx-lobes valvate.— Leaves pinnately compound or simple, or reduced to the petiole. Placenta 1, parietal. .... **Leguminosae**
1621. Leaves simple, undivided. Ovules 4–6, initially parietal, later central.—S.W. U.S., Mexico. .... **Fouquieriaceae**

— Leaves pinnately compound. Ovules 2, basal or parietal. **Connaraceae**
1622. Autotrophic, woody plants. Leaves well-developed........... 1623
— Saprophytic herbs. Leaves scale-like, not green.—Ovary 4- or 5-locular. Ovules many per locule.................. **Monotropaceae**
1623. Bark inside without tough, silky fibres. Ovary usually 3–20-locular, rarely 2-locular, then leaves small and endosperm present, or ovaries 2–more, free at base but not at the apex, then leaves translucent-glandular-punctate and ovules 2 per locule. 1624
— Bark inside with tough, silky fibres. Ovary 2-locular.—Leaves rather large to large. Flowers in umbels or in capitules. Anthers with 2 longitudinal slits. Ovule 1 per locule. Endosperm absent. (incl. *Aquilariaceae*). .... **Thymelaeaceae**
1624. Leaves simple, undivided.—Filaments free, rarely connate, then leaves small, narrow and usually in whorls. 1625
— Leaves pinnately compound, rarely simple and undivided, then rather large and filaments nearly completely connate, leaves alternate, rarely opposite. 1628
1625. Flowers bisexual. 1626
— Flowers unisexual or polygamous.—Flowers in racemes, 5-merous. Sepals free. Ovary stipitate. Ovules many per locule. ..**Capparaceae**
1626. Sepals usually connate. 1627
— Sepals free.—Flowers in racemes. Corolla imbricate. Anthers with terminal pores. Ovule 1 per locule. .... **Cyrillaceae**
1627. Leaves not translucent-glandular-punctate. Corolla imbricate, rarely valvate, then ovules 3–more per locule. Ovary 1.—Ovules 2–more per locule, rarely only 1, then flowers 4-merous. Endosperm copious. .... **Ericaceae**
— Leaves translucent-glandular-punctate. Corolla valvate. Ovaries 2–more, free at base, but not at the apex.—Flowers solitary or in fascicles. Ovules 2 per ovary......................... **Rutaceae**
1628. Twigs and petioles with a pale, wavy, sclerenchymatous ring around resinous ducts in transverse section. Filaments free. Ovules 2 per locule. .... **Burseraceae**
— Twigs and petioles without such a ring and ducts. Filaments nearly completely connate, rarely free, then ovules many per locule.—Anthers with 2 longitudinal slits...................... **Meliaceae**
1629. (1613). Ovary 1-locular. Ovules numerous.—Placenta parietal or basal-parietal. 1630
— Ovary 1, 2–more-locular, or ovaries 2–5. Ovules either few or axillary... 1631
1630. Leaves opposite. Sepals connate, valvate. Stamens 6. Ovules parietal. Endosperm present. .... **Frankeniaceae**
— Leaves alternate. Sepals free, imbricate. Stamens 8–10. Ovules

basal-parietal. Endosperm absent. (*Tamariceae*). ..... **Tamaricaceae**
1631. Ovary 1, undivided or lobed. Endosperm present. 1632
— Ovaries 2–5, free, or connate at base only. Endosperm absent.—
Woody plants. Leaves compound. Flowers in racemes or in panicles.
Stamens 10. Ovules 2 per ovary, collateral. .... **Connaraceae**
1632. Leaves simple.—Sepals connate. 1633
— Leaves compound, rarely unifoliolate.—Usually herbs. Stamens 10,
connate at base. Styles 5. (incl. *Averrhoaceae*). .... **Oxalidaceae**
1633. Stipules present. Flowers in panicles. Disk present.—Flowers unisexual. Ovule 1 per locule. .... **Euphorbiaceae**
— Stipules absent. Flowers in fascicles, or in cymes, or solitary. Disk
absent.—Woody plants. Leaves simple. Ovules 1 or 2 per locule.

Ebenaceae
1634. (1612). Style 1 *per flower*, simple, stigma 1, or 2–more, then adjacent at base. Ovary 1. 1635
— Styles 2–more *per flower*, free, or connate at base but not up to the
stigmas, or ovaries free, 3–more. 1646
1635. Ovary 1-locular. 1636
— Ovary at least in the older flowers 2–12-locular.—Woody plants.
Leaves simple, undivided. 1641
1636. Ovule or seed 1.—Filaments in bundles. 1637
— Ovules 2–more. 1638
1637. Calyx indistinct, at best consisting of tubercles. Anthers inserted on
a usually glandular hypanthium.—Corolla (in fact the single
perianth) well-developed. (*Pisonia*). .... **Nyctaginaceae**
— Calyx larger than the corolla, accrescent in fruit. Anthers apparently
basally attached. (*Pentaplaris*, ? misplaced in:). .... **Tiliaceae**
1638. Leaves simple, undivided. 1639
— Leaves pinnately compound or reduced to a broadened petiole.—
Stipules usually present. Corolla valvate. Stigma 1. Ovules parietal.
Endosperm scanty or absent. .... **Leguminosae**
1639. Sepals 3. Stigma 1. 1640
— Sepals 5. Stigmas 3 or 4.—Anthers with longitudinal slits. Ovules
many. Placentas several, initially parietal, later apparently axillary.
Subtropical N. America. .... **Fouquieriaceae**
1640. Filaments connate. Anthers with 1 slit. Stigma small. Ovules 6–8.
Placentas several, parietal.—Madagascar. (*Cinnamosma*).

Canellaceae
— Filaments free. Anthers with 2 longitudinal slits. Stigma broad.
Ovules many, irregularily placed on the wall. .... **Annonaceae**
1641. Petals connate at base only.—Petals imbricate or scale-like. 1642
— Petals completely connate.—Ovary 3–12-locular, initially incom-
pletely so. 1644

1642. Stipules present.—Sepals 5. Ovary 3-locular. Ovules 2 per locule.
Dipterocarpaceae
— Stipules absent.. 1643
1643. Ovary (8–)10–12-locular. Ovules solitary, pendulous.—Petals scale-like. Stamens free or connate in bundles. New Caledonia, Queensland, ?New Hebrides. (*Aquilariaceae*: *Lethedon*).....**Thymelaeaceae**
— Ovary either 5-locular with 2 ovules per locule, or 2–5-locular with many ovules per locule............................... **Theaceae**
1644. All flowers fertile, bracts not both coloured and saccate. Sepals connate, at least in bud. 1645
— Sterile flowers with coloured, saccate bracts. Sepals 4, free.—Stipules absent. Anthers with longitudinal slits. Tropical America.
Marcgraviaceae
1645. Stipules absent. Calyx persistent. Corolla-lobes entire. Anthers with terminal pores. Tropical Africa. .... **Scytopetalaceae**
— Stipules present. Sepal-lobes connate in bud, ultimately becoming free and deciduous. Corolla-lobes fimbriate. Anthers with lateral slits. New Guinea, New Caledonia. (*Antholoma*). .. **Elaeocarpaceae**
1646. (1634). Leaves simple.................................... 1647
— Leaves pinnately compound.—Stipules present. Corolla-lobes 4 or 5, valvate. Ovules many per carpel................. **Leguminosae**
1647. Stipules absent... 1648
— Stipules present.—Flowers unisexual. Corolla-lobes 5 or 6. Disk present. Ovary 2–4-locular. Ovule 1 per locule...... **Euphorbiaceae**
1648. Ovary 1, undivided or lobed.............................. 1650
— Ovaries 3 or more, free.—Leaves simple. Stipules absent. 1649
1649. Sepals 2 or 3. Corolla-lobes 3–6. .... **Annonaceae**
— Calyx and corolla both calyptrate, irregularily dehiscent at the base.—Vegetative parts covered with peltate scales. E. Malesia, N. Australia, W. Pacific........................ **Himantandraceae**
1650. Ovary 2–16-locular....................................... 1652
— Ovary 1-locular.—Ovules many. Placentas initially parietal, later apparently axillary....................................... 1651
1651. Leaves alternate or in fascicles. Sepals 5, free. Stamens 10–15.—Flowers in racemes or in panicles. Disk present. Endosperm scanty. Subtropical N. America. .... **Fouquieriaceae**
— Leaves opposite. Calyx tubular, 5- or 6-dentate. Stamens 20–more.—Inflorescence cymose. Halophylous plants.... **Frankeniaceae**
1652. Ovules either 2 per locule or many........................ 1653
— Ovules 1 or 2 per locule.—Flowers solitary, or in fascicles, or in cymes. Sepals connate. Disk absent. Ovules pendulous. Endosperm copious. .... **Ebenaceae**
1653. Corolla imbricate..................................... **Theaceae**

151

— Corolla contort. (*Bonnetiaceae*)........................ **Theaceae**

1654. (1574). Fertile stamens less than the corolla-lobes............ 1655

— Fertile stamens as many as the corolla-lobes or more......... 1680

1655. Stamens 2 – more.. 1656

— Stamen 1. 1658

1656. Stamens 5 – 16.—Corolla-lobes 10 or 15 – 24. 1657

— Stamens 2 – 4.—Corolla-lobes 3 – 12. 1660

1657. Corolla-lobes 10. Stamens 5. Ovary 1-locular. Ovules many.—Tropical America, West Indies. .... **Theophrastaceae**

— Corolla-lobes 15 – 24. Stamens 5 – 16. Ovary 5 – 12-locular. Ovule 1 per locule. .... **Sapotaceae**

1658. Leaves opposite. Corolla-lobes 4. Ovules many.............. 1659

— Leaves alternate. Corolla-lobes 5. Ovule 1 per locule.—Ovary 2 – 4-locular, lobed. .... **Boraginaceae**

1659. Herbs. Corolla imbricate. Ovary 1-locular............ **Gentianaceae**

— Lianas. Corolla valvate. Ovary 2-locular.—Tropical W. Africa. (*Antoniaceae*: *Usteria*). .... **Loganiaceae**

1660. Fertile stamens 2 (rarely 3), alternating with the locules of the ovary. 1661

— Fertile stamens 2 (rarely up to 4), not distinctly alternating with the locules of the ovary.—Disk usually present.................. 1662

1661. Leaves usually opposite. Disk absent. Anthers with 2 longitudinal slits.—Trees, shrubs, or undershrubs. Ovary 2- (rarely 3-) locular.
Oleaceae

— Leaves alternate. Disk present. Anthers with 1 longitudinal slit.— Undershrubs. Disk 4-partite. .... **Epacridaceae**

1662. Leaves opposite or in whorls. 1663

— Leaves alternate. 1672

1663. Ovule 1 per complete or incomplete locule.................. 1664

— Ovules 2 – more per locule. 1666

1664. Plants usually woody. Ovary completely or incompletely 2-, or 4-, rarely 8-locular. Fruit usually a drupe...................... 1665

— Plants usually herbaceous, occasionally undershrubs. Ovary 4-locular. Fruit usually dehiscing into 4 drupelets. .... **Labiatae**

1665. Endosperm present.—Ericoid undershrubs. Leaves narrow, in whorls. Spikes racemose. Ovules basal, apotropous. S. Africa. (*Stilbaceae*).. **Verbenaceae**

— Endosperm absent, if present flowers solitary or in cymose inflorescences and ovule either axillary and campylotropous, or apical and atropous. .... **Verbenaceae**

1666. Ovary 2-, rarely 1-locular. 1667

— Ovary 5-locular.—Leaves translucent-glandular-punctate. Ovules 2 per locule. .... **Rutaceae**

1667. Seeds on enlarged, indurated, more or less hook-shaped funicles (retinacula), rarely without these, then either sepals connate at base only, or ovules (and seeds) 1 or 2 per locule.—Leaves simple. Fruit a loculicid capsule, placentas persisting on the valves, rarely a 1- or 2-seeded drupe.. 1668

— Seeds without retinacula. 1670

1668. Fruit a 2–many-seeded capsule. Ovary 2-locular.............. 1669

— Fruit a 1- or 2-seeded drupe. Ovary 1-, rarely 2-locular. (*Mendonciaceae*). .... **Acanthaceae**

1669. Retinacula well-developed. Ovules 1–many per locule. **Acanthaceae**

— Retinacula absent or papillate. Ovules 2 per locule. (*Thunbergiaceae*). .... **Acanthaceae**

1670. Leaves simple, in aquatic herbs the submerged ones dissected. 1671

— Leaves usually compound.—Sepals nearly completely connate, if only at base then leaves compound. Stigmas 2. Fruit a septifragous or loculicide capsule, placentas persisting on the enlarged sept. Seeds 2–many per locule, usually winged. Endosperm absent.

Bignoniaceae

1671. Stamens 4. Fruit a capsule. Endosperm present. (*Bacopa, Freylinia*)....................................... **Scrophulariaceae**

— Stamens 2. Fruit a berry, usually white. Endosperm absent.—Malesia to Polynesia. (*Cyrtandra*)..................... **Gesneriaceae**

1672. (1662). Ovary 1-locular. Ovules 3–7.—Stamens 3............ 1673

— Ovary 2–10-locular, rarely 1-locular, then ovule 1. 1674

1673. Trees, shrubs, or undershrubs. Calyx cupuliform, 3–6-dentate. Ovules 3. .... **Olacaceae**

— Cushion-forming perennials. Sepals 2, free. Ovules 4–7. New Zealand. (*Hectorella*). .... **Hectorellaceae**

1674. Ovary 2–10-locular. Ovules 1 or 2 per locule................ 1675

— Ovary 2-locular, ovules many per locule, rarely 1-locular and ovule 1. 1678

1675. Corolla-lobes 4.—Herbs, sometimes woody at base. Flowers in spikes or in capitules, rarely solitary. Stamens 2. Ovary 2–4-locular. Fruit a capsule with a lid......................... **Plantaginaceae**

— Corolla-lobes 5.. 1676

1676. Stem woody. Ovules pendulous.—Leaves often translucent-glandular-punctate....................................... 1677

— Herbs. Ovules ascending.—Leaves not translucent-glandular-punctate. Ovary 2–4-locular, lobed. Ovule 1 per locule. Fruit a schizocarp or a capsule........................... **Boraginaceae**

1677. Fertile stamens 4. Fruit a drupe or a nut.—Leaves undivided, often translucent-glandular-punctate. Anthers confluent at the apex. Disk obscure to absent................................ **Myoporaceae**

153

— Fertile stamens 2 or 3. Fruit a loculicide and septicide capsule.—
Leaves translucent-glandular-punctate. Ovary 5-locular. Ovules 2
per locule. .... **Rutaceae**
1678. Corolla valvate or plicate, then sometimes imbricate. Sept of the
ovary usually oblique to the plane of symmetry of the flower.—
Leaves alternate, sometimes paired, but not opposite. Flowers solitary or in cymes. Fruit a septicid capsule, rarely a berry. Endosperm
present. .... **Solanaceae**
— Corolla imbricate, not plicate, rarely valvate or plicate, then leaves
opposite. Sept of the ovary usually at a right angle to the plane of
symmetry of the flower. 1679
1679. Seeds few, peltate, minutely pubescent.—Prostrate herbs. Leaves
alternate, pinnatifid. Flowers solitary. Disk large, cupular. Capsule
stipitate. India to New Guinea. (*Ellisiophyllaceae*, sometimes in
Hydrophyllaceae)............................. **Scrophulariaceae**
— Seeds many, not peltate, glabrous. Plants otherwise.

Scrophulariaceae
1680. (1654). Fertile stamens as many as the corolla-lobes........... 1681
— Fertile stamens more than the corolla-lobes. 1859
1681. Stamens alternipetalous. 1682
— Stamens epipetalous. 1839
1682. Style either 1, or 1 per ovary when ovaries free, simple with 1 or
2 – more stigmas adjacent at base, or absent and stigma 1, sessile.

1683
— Styles 2 – more, free or connate at base but not up to the stigmas, or
connate at the apex only, or stigmas 2 – more, sessile. Ovaries when
apparently free with common styles or stigmas................ 1812
1683. Ovary 1, 1-locular, sometimes incompletely so................ 1684
— Ovary 1, 2 – more-locular or nearly so, or ovaries 2 – more, free.1711
1684. Ovule 1.. 1685
— Ovules 2 – more... 1688
1685. Ovule basal... 1686
— Ovule apical.—Stipules absent. Flowers 5-merous............. 1687
1686. Woody plants. Leaves opposite. Stipules minute. Flowers 4-merous.
Corolla imbricate............................. **Salvadoraceae**
— Herbs. Leaves radical. Stipules absent. Flowers 5-merous. Corolla
valvate.—Flowers in a capitule, almost actinomorphic. Anthers connate. Stigma with a cup-shaped involucre. Australia... **Brunoniaceae**
1687. Ericoid shrubs. Leaves fan-nerved, white underneath, less than 2.5
cm long. Anthers with 1 longitudinal slit.—Leaves alternate. Australia. (*Monotoca*)............................. **Epacridaceae**
— Shrubs, sometimes climbing with tendrils. Leaves pinninerved, not
conspicuously white underneath and larger. Anthers with 2 longi-

tudinal slits. N.W. S. America. (*Metteniusa*, also in *Icacinaceae*, in *Opiliaceae* as *Aveledoa*)............................ **Alangiaceae**

1688. Ovules 2 or 3.. 1689
— Ovules 4 – more.. 1696

1689. Ovules 3.—Woody plants. Leaves alternate. Corolla 5-partite, valvate... 1690
— Ovules 2.. 1691

1690. Flowers in racemes. Filaments connate at base. Fruit a drupe.

Styracaceae

— Flowers solitary, also from the axils of fallen leaves. Filaments inserted on the corolla. Fruit a berry.—Mexico. (*Goetzeaceae*: *Lithophytum*).. **Solanaceae**

1691. Leaves simple, undivided, or lobed.—Stipules absent. Calyx apert or imbricate.. 1692
— Leaves pinnately compound, rarely reduced to the petiole or absent, then stipules present.................................... 1695

1692. Woody plants or twining herbs. Corolla valvate or imbricate. Ovules pendulous.—Fruit indehiscent. 1693
— Herbs, usually climbing with tendrils or prostrate. Corolla plicate. Ovules erect................................. **Convolvulaceae**

1693. Erect plants, rarely climbing. Bark without white juice. 1694
— Herbaceous climbers with abundant white juice.—Stipules absent. Inflorescence cymose, cincinnoid. Calyx and corolla imbricate. Ovules apical, pendulous. Fruit dry, indehiscent, winged. S.E. Asia to Australia................................. **Cardiopteridaceae**

1694. Ovules pendulous.—Erect plants, rarely climbing. ...... **Icacinaceae**
— Ovules basal.—Erect shrubs. Mexico. (*Goetzeaceae*: *Lithophytum*).

Solanaceae

1695. Stipules usually present. Flowers in spikes or in capitules. Calyx and corolla valvate. Ovules serial. Placenta 1, parietal. Fruit a dehiscent pod.—Leaves rarely reduced to the petiole or absent.. **Leguminosae**
— Stipules absent. Flowers in a panicle. Aestivation various. Ovules not serial on 1 placenta. Fruit a berry.—Petals connate at base only. Filaments nearly completely connate.................. **Meliaceae**

1696. Ovules 4. 1697
— Ovules 5 – more... 1701

1697. Leaves alternate.—Ovules basal. 1698
— Leaves opposite or in whorls.—Corolla imbricate............. 1700

1698. Plants not cushion-forming. Leaves well-developed, distant. Sepals 5. Disk present... 1699
— Cushion-forming perennials with densely imbricate, small leaves. Sepals 2. Corolla valvate. Disk absent. New Zealand. (*Hectorella*).

Hectorellaceae

1699. Plants usually climbing with tendrils. Corolla induplicative-plicate.

                                                Convolvulaceae

 — Plants erect. Corolla valvate.—Mexico. (*Goetzeaceae: Lithophytum*).. **Solanaceae**

1700. Sepals connate. Endosperm absent. (incl. *Avicenniaceae, Symphoremataceae*). .... **Verbenaceae**

 — Sepals free. Endosperm usually present.—Woody plants. Flowers 5-merous. Fruit a capsule. .... **Celastraceae**

1701. Ovules basal or central................................. 1702

 — Ovules parietal...................................... 1705

1702. Corolla valvate...................................... 1703

 — Corolla imbricate.................................... 1704

1703. Woody plants. Leaves well-developed, distant, opposite. Calyx 4- or 5-merous. Fruit a berry. (*Strychnaceae: Strychnos*)..... **Loganiaceae**

 — Cushion-forming perennials with densely imbricate, small leaves. Sepals 2. Fruit a capsule. New Zealand. (*Hectorella*). **Hectorellaceae**

1704. Woody plants. Stipules absent. Fruit a drupe.......... **Verbenaceae**

 — Herbs. Stipules present. Fruit a capsule.—Sepals free. Endosperm present. Embryo curved...................... **Caryophyllaceae**

1705. Placentas 2 – more...................................... 1706

 — Placenta 1.—Leaves alternate, pinnately compound, rarely reduced to a broadened petiole or absent. Stipules usually present. Corolla valvate. (*Mimosoideae*). .... **Leguminosae**

1706. Leaves simple, rarely digitately compound. 1707

 — Leaves pinnately compound.—Woody plants. Leaves usually alternate. Stipules absent. Corolla valvate. Filaments nearly completely connate. Ovary initially 4- or 5-locular. .... **Meliaceae**

1707. Apex of the style stigmatic on the lower or outer side of a thickened part, summit glabrous.—Woody plants. Latex present. Flowers 5-merous.. **Apocynaceae**

 — Stigma apical on the style, or up to it, or between its lobes. Latex absent. 1708

1708. Corolla valvate.—Endosperm present...................... 1709

 — Corolla imbricate or contort.............................. 1710

1709. Shrubs or trees. Leaves opposite. (*Strychnaceae: Strychnos*).

                                                Loganiaceae

 — Herbs. Leaves radical or alternate. (*Menyanthaceae*). . **Gentianaceae**

1710. Corolla contort, rarely imbricate. Fruit a septicide capsule or a berry. Endosperm present.—Sap bitter. .... **Gentianaceae**

 — Corolla imbricate. Fruit a loculicide capsule. Endosperm absent.—Herbs or shrublets. Leaves usually radical........... **Gesneriaceae**

1711. (1683). Ovary 2-locular, or ovaries 2, free................... 1712

 — Ovary 3 – more-locular, or ovaries 3 – more, free............. 1759

156

1712. Ovule 1 per locule or free ovary. 1713
— Ovules 2 – more per locule or free ovary. 1725
1713. Leaves simple, sometimes dissected. 1714
— Leaves pinnately compound.—Woody plants. Filaments nearly completely connate. . . . . **Meliaceae**
1714. Leaves all, or only the upper opposite.—Style apically stigmatic.
1715
— Leaves all, or only the upper alternate, or all radical. 1719
1715. Stipules absent or reduced to an interpetiolary line. 1716
— Stipules present.—Stem woody. Leaves undivided. Calyx shallowly lobed. Corolla valvate. Disk absent. Stigmas 2. Ovules erect. Fruit a drupe. Endosperm present. . . . . **Rubiaceae**
1716. Corolla imbricate. Ovule basal, or apical, or axillary, then plant herbaceous or ovule campylotropous. Fruit a drupe, or a schizocarp, or a capsule. 1717
— Corolla valvate. Ovule axillary, hemitropous. Fruit a berry.— Woody plants. Flowers in cymose panicles. Endosperm present. (incl. *Strychnaceae*). . . . . **Loganiaceae**
1717. Fruit a 2-valved capsule, or a drupe, or a schizocarp. Endosperm absent.—Corolla coloured. 1718
— Fruit a capsule with a lid. Endosperm present.—Herbs. Flowers in spikes or capitules. Corolla membranous. Stigma undivided. Ovule axillary, hemitropous. (*Plantago*). . . . . **Plantaginaceae**
1718. Ovule apical or axillary, anatropous or campylotropous. Fruit a drupe or a schizocarp. . . . . **Verbenaceae**
— Ovule apical, atropous. Fruit a 2-valved capsule.—Woody mangrove plants with pneumatophores (adventitious roots sticking up out of the mud). Flowers in dense, leafy, cymose spikes, 4-merous. (*Avicenniaceae*). . . . . **Verbenaceae**
1719. Flowers 5-merous. 1720
— Flowers 4-merous.—Herbs or undershrubs, rarely shrubs. Flowers in spikes or in capitules, rarely solitary. Sepals connate at base. Disk absent. Anthers with 2 longitudinal slits. Fruit dehiscing with a lid.
Plantaginaceae
1720. Sepals connate. Anthers with 2 longitudinal slits. 1721
— Sepals free or nearly so. Anthers with 1 longitudinal slit.—Shrubs or trees, rarely undershrubs. Fruit a berry or a drupe. . . . . **Epacridaceae**
1721. Corolla imbricate. 1722
— Corolla plicate. 1724
1722. Herbs. Corolla-tube about as long as the limb. Fruit a capsule or a schizocarp. Embryo straight. 1723
— Large trees. Corolla-tube much longer than the limb. Fruit a drupe. Embryo horse-shoe-shaped.—Flowers in cymes. Style apical. Stigma

1. Amazonia. (sometimes included in *Apocynaceae* or *Boraginaceae*)................................... **Duckeodendraceae**
1723. Style apical. Stigmas 2. Fruit a capsule.—Flowers usually in capitules, very small. Temperate America. (*Collomia*) .... **Polemoniaceae**
— Style gynobasic. Stigma 1. Fruit a schizocarp.—Flowers in cincinni. Temperate Old World and Australia. (*Rochelia*)...... **Boraginaceae**
1724. Calyx dentate....................................... **Solanaceae**
— Calyx divided more deeply. .... **Convolvulaceae**
1725. (1712). Ovules 2 per locule or free ovary.................... 1726
— Ovules 3 – more per locule or free ovary. 1741
1726. Style stigmatic on the apex, or up to it, or between its lobes. Ovary 1, undivided or shallowly lobed. 1727
— Apex of the style stigmatic on the lower or the outer side of a thickened part, summit glabrous.—Stem woody. Latex present. Leaves undivided, usually opposite. Stipules absent. ... **Apocynaceae**
1727. Leaves all or only the upper opposite or in whorls. 1728
— Leaves all or only the upper alternate, or all radical, or absent. 1731
1728. Disk absent.—Woody plants....................... 1729
— Disk usually present.—Sepals connate. Micropyle and radicle pointing down. 1737
1729. Sepals connate. Corolla with a distinct tube. Fruit a drupe..... 1730
— Sepals free. Petals connate at base only. Fruit dehiscent.—Sepals 5. Fruit 1-locular, 1-seeded. Endosperm present......... **Celastraceae**
1730. Climbing shrubs. Flowers in involucrate capitules; bracts 6. Stamens 5 – 16. Ovules anatropous. (*Symphoremataceae*). ...... **Verbenaceae**
— Erect shrubs or trees. Flowers in racemes. Stamens 4. Ovules anatropous. .... **Oleaceae**
1731. Ovules either basal, or ascending, or patent, or axillary and hemitropous. 1732
— Ovules pendulous, anatropous, raphe ventral................. 1739
1732. Ovules basal................................... 1733
— Ovules axillary.—Sepals connate. Corolla imbricate. Endosperm present. Embryo straight. 1735
1733. Trees or shrubs. Corolla valvate. Endosperm absent. Embryo straight or slightly curved.—Calyx 4 – 6-lobed, lobes valvate. West Indies. (*Goetzeaceae*: *Coeloneurum*, *Goetzea*). .... **Solanaceae**
— Climbers or twiners, rarely erect plants. Corolla plicate, or induplicate, or imbricate. Endosperm present. Embryo curved or folded. 1734
1734. Twining parasites. Flowers 4-merous. Calyx connate. Corolla imbricate. (*Cuscutaceae*)............................... **Convolvulaceae**
— Climbers, rarely erect plants, not parasiting. Flowers 5-merous. Sepals usually nearly free. Corolla plicate or induplicate. ...... **Convolvulaceae**
1735. Flowers 5-merous, rarely 4-merous, then stigmas 2. Disk more or

less developed, hypogynous. Fruit a capsule or a drupe........ 1736

— Flowers 4-merous. Fruit dehiscing with a lid.—Leaves undivided or lobed. Flowers in spikes, or in capitules, rarely solitary and terminal. Stigma 1. Ovules axillary, campylotropous. .. **Plantaginaceae**

1736. Herbs. Fruit a capsule. Ovules apotropous......... **Polemoniaceae**

— Shrubs. Fruit a drupe. Ovules epitropous.—Pantropical, restricted to riverbeds. (*Ehretiaceae: Rotula*)................ **Boraginaceae**

1737. Ovary completely 2-locular. Fruit a septicide capsule.—Disk present... 1738

— Ovary more or less incompletely loculed. Fruit a schizocarp or a drupe.—Funicles inconspicuous, seeds (sub-)sessile. Endosperm absent... **Verbenaceae**

1738. Ascending herb. Leaves opposite. Calyx deeply 3-fid. Funicles large, indurated, hook-shaped. Endosperm absent. Brazil. (*Pentstemonacanthus*)... **Acanthaceae**

— Virgate undershrubs. Leaves in whorls. Colyx 5-dentate. Funicles inconspicuous, seeds (sub-)sessile. Endosperm present. S. Africa. (*Retziaceae*, also in *Scrophulariaceae*, *Solanaceae*). .... **Loganiaceae**

1739. Styles 2. Stigma 1 per style.—Ericoid shrublets. Stipules absent. Sepals connate. Corolla imbricate, not plicate. Fruit a capsule or a nut. Endosperm present. Embryo minuscule. S. Africa. . **Bruniaceae**

— Style 1. Stigmas 1 or 2. 1740

1740. Leaves undivided. Stipules present, though often inconspicuous. Flowers in cymes. Stigmas 2.—Flowers 5-merous. Ovules pendulous.
Dichapetalaceae

— Leaves usually pinnately compound. Stipules absent. Flowers in panicles. Stigma 1, undivided or lobed.—Filaments nearly completely connate.................................... **Meliaceae**

1741. (1725). Stigma apical on the style or immediately below it, or between its apical lobes.—Ovary 1, undivided or shallowly lobed. 1742

— Apex of the style stigmatic on the lower or outer side of a thickened part, summit glabrous.—Latex present. Leaves undivided, usually opposite. Stipules absent. Flowers 5-merous. Ovary 1, 2-locular.
Apocynaceae

1742. Leaves all, or only the upper, alternate, or all radical. 1743

— Leaves all, or only the upper, opposite or in whorls.......... 1746

1743. Bracts, if any, not transformed. 1744

— Bracts of sterile flowers saccate, pitcher-like, or spathulate, brightly coloured.—Woody plants. Leaves simple. Flowers 5-merous. Sepals free. Corolla imbricate, 5-partite. Tropical America. **Marcgraviaceae**

1744. Sepals connate... 1745

— Sepals free.—Trees. Corolla unilaterally induplicate. Flowers solitary, axillary. Madagascar. (*Humbertiaceae*)......... **Convolvulaceae**

1745. Leaves simple.. 1753

— Leaves 1- or 3-foliolate.—Woody plants. Corolla nearly actinomorphic, 5-lobed, imbricate. Endosperm absent. ..... **Bignoniaceae**

1746. Fruit either septicide, or both septicide and loculicide, or indehiscent. Funicles not indurated, seeds (sub-)sessile. Endosperm present. 1747

— Fruit loculicide. Funicles indurated, more or less hook-shaped. Endosperm absent.—Stipules absent. Corolla imbricate, often contort.
Acanthaceae

1747. Leaves pinnately compound.—Trees (*Oroxylum*) or lianas (*Nyctocalos*). Corolla imbricate. S.E. Asia. .... **Bignoniaceae**

— Leaves simple... 1748

1748. Sap bitter. Corolla contort, segments overlapping to the right (laterally seen).—Herbs or undershrubs, rarely shrubs. Stipules absent,
sometimes an interpetiolary line present. .... **Gentianaceae**

— Sap not bitter. Corolla either valvate, or imbricate but not contort,
or contort and segments overlapping to the left, rarely to the right,
then plants woody and leaves either with a sheath at base, or
auriculate. 1749

1749. Corolla valvate... 1750

— Corolla imbricate or contort............................... 1752

1750. Woody plants. Corolla valvate or induplicative-valvate. 1751

— Herbs. Corolla exduplicative-valvate.—Style articulated. (*Spigeliaceae: Spigelia*)..................................... **Loganiaceae**

1751. Virgate shrub, glandular-hairy. Leaves in whorls. Stipules absent. Corolla induplicative-valvate.—S. Africa. (*Retziaceae*, also in
Scrophulariaceae or *Solanaceae*)..................... **Loganiaceae**

— Shrubs or trees, not glandular-hairy. Leaves opposite. Stipules connate into a sheath, or reduced to an interpetiolary line. Corolla
valvate. (*Antoniaceae, Strychnaceae*). .... **Loganiaceae**

1752. Corolla imbricate, rarely contorted to the left. Fruit a capsule.
Loganiaceae

— Corolla contorted to the right. Fruit a berry. (*Potaliaceae*). **Loganiaceae**

1753. Calyx 5-merous. 1754

— Calyx 4-partite.—Herbs or undershrubs. Flowers in spikes, or in
capitules, rarely solitary, terminal. Corolla 4-lobed, imbricate, not
plicate. Disk absent. Stigma undivided. Capsule dehiscing with a lid.
(*Plantago*)...................................... **Plantaginaceae**

1754. Corolla-lobes 5, imbricate, not plicate. Fruit a capsule, dehiscing
longitudinally.—Plants usually herbaceous.................... 1756

— Corolla valvate or plicate, then sometimes imbricate, rarely imbricate and not plicate, then plants *either* herbaceous with undivided or
lobed leaves, flowers solitary and axillary, calyx actinomorphic, 5-

160

fid, corolla 5-lobed, and stigma undivided or lobed, *or* plants woody and flowers solitary and axillary, *or* flowers in fascicles or cymose racemes and corolla 5-, rarely 4-lobed or -partite. Fruit a berry or a capsule dehiscing with a lid.................................. 1755

1755. Ovules 4. Endosperm absent.—West Indies. (*Goetzeaceae: Goetzea*). .... **Solanaceae**

— Ovules usually numerous, rarely 4. Endosperm present. (incl. *Salpiglossidaceae*). .... **Solanaceae**

1756. Leaves not both filiform and circinnate in bud. Corolla-lobes imbricate, distinctly connate. 1757

— Leaves filiform, circinnate in bud. Corolla-lobes contort, nearly free to base.—Herbs, sometimes slightly shrubby with glandular hairs (insectivorous). Flowers solitary, axillary. Capsule 2–4-valved. Australia.. **Byblidaceae**

1757. Leaves pinninerved, entire to pinnate. Flowers not in secund cincinni. Stigmas 1 or 2..................................... 1758

— Leaves palmately lobed. Flowers in secund cincinni. Stigma 2-, rarely 3-lobed.—Ovules many per locule. (*Romanzoffia*).

Hydrophyllaceae

1758. Flowers minute, in capitules. Stigmas 2. Ovules rather few per locule. Fruit 3-valved. Testa mucilaginous. (*Collomia*).

Polemoniaceae

— Flowers relatively large, not in capitules. Stigma 1. Ovules many per locule. Fruit 2-valved, valves sometimes bifid. Testa not mucilaginous. (*Verbascum*). .... **Scrophulariaceae**

1759. (1711). Ovary 3-locular, or ovaries 3, free.................... 1760

— Ovary 4–more-locular, or ovaries 4–more, free.............. 1784

1760. Stigma 1, undivided or lobed. 1761

— Stigmas 3...................................... 1764

1761. Leaves opposite....................................... 1762

— Leaves alternate. 1772

1762. Stipules absent.. 1763

— Stipules present.— Disk absent. Ovary undivided. Fruit a capsule. (*Geniostoma*). .... **Loganiaceae**

1763. Flowers in cymes. Disk absent. Ovary bipartite. Fruit a berry.

Apocynaceae

— Flowers in umbels. Disk present. Ovary undivided. Fruit a capsule.

Ericaceae

1764. Corolla contort, not plicate.—Flowers 5-merous. Sepals connate. Disk present. Stigmas linear. Fruit dry..................... 1765

— Corolla valvately plicate or imbricate, not contort. 1766

1765. Woody climbers. Leaves paripinnate, the terminal pair of leaflets transformed into tendrils. Stipules usually resembling the lower

161

leaflets. Capsule septicide. (*Cobaeaceae*)............ **Polemoniaceae**
— Plants rarely woody, never climbing. Leaves, if compound, imparipinnate, without tendrils. Stipules absent. Fruit a loculicide capsule
or indehiscent.. **Polemoniaceae**
1766. Leaves alternate or in pairs but not opposite. 1767
— Leaves opposite.—Woody plants. Disk absent. Endosperm present.
1771
1767. Stipules absent. Endosperm present. 1768
— Stipules present. Endosperm absent. .... **Dichapetalaceae**
1768. Ovules many per locule. 1769
— Ovules 2 per locule.—Sepals free. Corolla plicate. Disk present.
Convolvulaceae
1769. Corolla imbricate. Disk absent........................... 1770
— Corolla valvate or plicate. Disk present.—Leaves alternate or in
pairs, but not opposite. Fruit a berry. .... **Solanaceae**
1770. Woody plants.. **Theaceae**
— Herbs.. **Diapensiaceae**
1771. Stipules absent. Sepals free. Corolla imbricate. Ovules 2 per locule.
Fruit a capsule.—Flowers 5-merous. S.E. Asia to China, Mexico, C.
America. (*Microtropis*). .... **Celastraceae**
— Stipules present. Sepals connate. Corolla valvate. Ovule 1 per
locule. Fruit a drupe............................... **Rubiaceae**
1772. Flowers 4-merous. 1773
— Flowers 5-merous. 1776
1773. Corolla imbricate.................................... 1774
— Corolla valvate or plicate. 1775
1774. Flowers in spikes or in capitules. Filaments free. Endosperm
present.—Herbs or undershrubs. Calyx deeply divided. Disk absent.
Fruit a capsule.............................. **Plantaginaceae**
— Flowers in panicles. Filaments nearly completely connate. Endosperm absent.—Ovules 1 or 2 per locule. .... **Meliaceae**
1775. Ovules 1 or 2 per locule. Endosperm absent.—Corolla valvate. Filaments nearly completely connate. Anthers with longitudinal slits.
Meliaceae
— Ovules many per locule. Endosperm present.—Disk present. Fruit a
berry. .... **Solanaceae**
1776. Corolla plicate. 1777
— Corolla valvate or imbricate, not plicate. 1779
1777. Sepals connate. Embryo curved........................... 1778
— Sepals free. Embryo plicate.—Herbs or undershrubs. Disk present.
Anthers with 2 longitudinal slits. Ovary undivided. Ovules 2 per
locule. Fruit a capsule. .... **Convolvulaceae**
1778. Ovary either undivided, ovules 3–6 per locule, or ovary deeply di-

vided, ovules 1 or 2 per locule. Fruit a schizocarp....... **Nolanaceae**

— Ovary undivided. Ovules many per locule. Fruit a berry. **Solanaceae**

1779. Corolla valvate.. 1780

— Corolla imbricate.—Anthers with 1 transverse or 2 longitudinal slits.
1783

1780. Filaments free. 1781

— Filaments connate.—Anthers with 2 longitudinal slits.......... 1782

1781. Anthers with 1 longitudinal slit. Ovule 1 per locule.—Shrubs or trees. Disk present. Fruit a drupe. .... **Epacridaceae**

— Anthers with 2 longitudinal slits or pores. Ovules many per locule.—Endosperm present......................... **Solanaceae**

1782. Flowers in panicles. Ovules 1 or 2 per locule. Endosperm absent.
Meliaceae

— Flowers in racemes. Ovule 1 per locule. Endosperm present.
Styracaceae

1783. Herbs or undershrubs. Ovules many per locule.—Disk absent. Fruit a capsule. .... **Diapensiaceae**

— Woody plants. Ovules 1 or 2 per locule................. **Meliaceae**

1784. (1759). Ovary 4-locular, or ovaries 4, free.................. 1785

— Ovary 5–more-locular, or ovaries 5–more, free............. 1786

1785. Style stigmatic at the apex or between the apical lobes. 1788

— Styles stigmatic below the apex, usually free at base.—Tropical Africa. (*Pleiocarpa*). .... **Apocynaceae**

1786. Leaves compound....................................... 1787

— Leaves simple... 1802

1787. Leaves digitately 3-foliolate, translucent-glandular-punctate. Filaments free....................................... **Rutaceae**

— Leaves pinnately compound, not translucent-glandular-punctate. Filaments nearly completely connate.................... **Meliaceae**

1788. Ovules 1 or 2 per locule, rarely more, then corolla imbricate, not plicate. 1789

— Ovules many per locule. Corolla plicate or valvate.—Leaves alternate or in pairs but not opposite. Sepals connate. Anthers with 2 longitudinal slits or apical pores .... **Solanaceae**

1789. Anthers with 2 longitudinal slits; thecae rarely apically confluent.
1790

— Anthers with 1 longitudinal slit.—Woody plants. Sepals free.
Epacridaceae

1790. Leaves opposite or in whorls, exceptionally alternate, then leaves simple, flowers solitary, axillary, corolla with a distinct tube, 4-merous, stigmas 2, fruit a drupe.......................... 1791

— Leaves alternate, at least the upper, or all radical............. 1795

1791. Ovary undivided or shallowly lobed. 1792

163

— Ovary deeply divided.—Ovule 1 per locule.................... 1793
1792. Ovule 1 per locule. (Endosperm present: *Dicrastylidaceae-Physopsideae*). (incl. *Avicenniaceae*)................. **Verbenaceae**
— Ovules many per locule. (*Potaliaceae*: *Anthocleista, Potalia; Buddlejaceae*: *Buddleja*)............................. **Loganiaceae**
1793. Flowers 5-merous, in cincinni.—Calyx divided. Stigma 1.

Boraginaceae
— Flowers 4-merous, solitary or in false whorls or in panicles..... 1794
1794. Flowers solitary. Ovule erect, atropous.—Creeping herbs, rooting at the nodes. New Zealand, Patagonia. (*Tetrachondraceae*, also in *Labiatae, Scrophulariaceae*)....................... **Boraginaceae**
— Flowers in false whorls or in panicles. Ovule anatropous, erect, apotropous. .... **Labiatae**
1795. Leaves simple... 1796
— Leaves pinnately compound.—Woody plants. Filaments nearly completely connate................................... **Meliaceae**
1796. Petals connate into a distinct tube, which is rarely very short, then either stem herbaceous, or anthers connate, or ovules many.... 1797
— Petals only slightly connate at base.—Stem woody. Anthers free. Corolla imbricate. Disk absent. Ovary undivided. Ovules 1 or 2 per locule. Fruit a drupe or a berry. .... **Aquifoliaceae**
1797. Flowers 5-merous, very rarely 4-merous, then stem woody, disk present, and fruit a drupe. 1798
— Flowers 4-merous.—Stem herbaceous or woody at base only. Flowers in spikes or in capitules, rarely solitary, terminal. Calyx divided. Corolla imbricate. Disk absent. Stigma 1. Fruit dehiscing with a lid. Embryo straight or nearly so, radicle pointing down. .... **Plantaginaceae**
1798. Ovules 1–3 per locule. 1799
— Ovules many per locule. 1801
1799. Corolla imbricate or contort, not plicate. 1800
— Corolla valvate or plicate.—Ovules 1 or 2 per locule. Micropyle and radicle pointing down. Embryo curved or plicate.   **Convolvulaceae**
1800. Style terminal. Ovules 2 or 3 per locule, apotropous. Fruit a berry or a capsule. .... **Solanaceae**
— Style usually gynobasic, rarely terminal, then plants woody, tropical, ovule 1 per locule, and fruit a drupe (*Ehretiaceae*: *Lepidocordia, Rotula*). Ovules 1 or 2 per locule, epitropous. Fruit a drupe or a schizocarp. .... **Boraginaceae**
1801. Flowers in terminal racemes. Bracts saccate, brightly coloured.— Tropical America........................... **Marcgraviaceae**
— Flowers solitary or in axillary racemes. Bracts not so. ...... **Theaceae**
1802. (1786). Corolla with a distinct tube........................ 1803
— Petals only slightly connate at base.—Shrubs or trees. Sepals con-

nate. Corolla imbricate. Anthers with 2 longitudinal slits. Disk absent. Ovary undivided. Ovules 1 or 2 per locule. Fruit a drupe or a berry. Embryo straight. . . . . **Aquifoliaceae**

1803. Corolla valvate or imbricate, not plicate. 1804
— Corolla plicate.—Flowers solitary. Sepals connate. Anthers with 2 longitudinal slits. Fruit a schizocarp. Embryo curved. W. S. America. . . . . **Nolanaceae**

1804. Plants autotrophic with green leaves. Ovary 5–10-locular. 1805
— Parasitic herbs. Leaves scale-like, brown. Ovary 12–28-locular.— Flowers in spikes, or in capitules, or in panicles. Sepals free. Corolla imbricate. Disk absent. Anthers with 2 longitudinal slits. Ovule 1 per locule. S.W. U.S., Mexico. . . . . **Lennoaceae**

1805. Anthers with 1 transverse or 2 longitudinal slits. 1806
— Anthers with 1 longitudinal slit.—Leaves alternate, rarely opposite, then ovule 1 per locule. Sepals free or nearly so. Disk usually present. . . . . **Epacridaceae**

1806. Anthers with 2 longitudinal not confluent slits. 1807
— Anthers with 1 transverse or 2 confluent slits.—Woody plants. Leaves alternate. Sepals connate. Ovary simple. Ovule 1 per locule.
Myoporaceae

1807. Leaves alternate. 1808
— Leaves opposite.—Woody plants. Sepals connate. 1810

1808. Sepals connate. Ovules 1 or 2 per locule. 1809
— Sepals free. Ovules many per locule.—Bracts saccate, brightly coloured. Ovary undivided, 5- or 6-locular. Tropical America.
Marcgraviaceae

1809. Herbs. Leaves not translucent-glandular-punctate. Ovary 10-locular. Ovule 1 per locule. . . . . **Boraginaceae**
— Woody plants. Leaves translucent-glandular-punctate. Ovary 5-partite. Ovules 2 per locule. . . . . **Rutaceae**

1810. Ovules 1 or 2 per locule or free ovary. 1811
— Ovules many per locule.—Leaves spinous. Ovary undivided, 5-locular at base, 1-locular at the apex. Andes. (*Potaliaceae*: *Desfontainia*). . . . . **Loganiaceae**

1811. Ovary undivided. . . . . **Verbenaceae**
— Ovaries 5, free.—Tropical Africa. (*Pleiocarpa*). . . . . **Apocynaceae**

1812. (1682). Styles free at base, connate at the more or less thickened apex.—Leaves usually opposite. Styles 2(–5). 1813
— Styles entirely free, or connate at base only. 1816

1813. Flowers 5-merous. Style with a thickened apex stigmatic on its sides or base. 1814
— Flowers 4-merous. Style apically stigmatic.—Herbs. Leaves at base with an interpetiolary ridge or sheath. Ovary 2-locular.

(*Spigeliaceae*: *Mitrasacme*). .... **Loganiaceae**

1814. Stamens free. Pollen free.................................. 1815

— Stamens connate and adnate to the style apex into a ± capitate
body. Pollen coherent into pollinia.—Leaves above with or without
a tuft of short, cylindric, hair-like appendages ('colleters') at the base
of the midrib. Corolla often more or less urceolate, the tube usually
shorter than the lobes............................ **Asclepiadaceae**

1815. Anthers coherent and appressed against the apex of the style, alternating with spathulate appendages of the latter on which the pollen
is discharged and which conceal the stigmatic areas of it. (*Periplo-
caceae*)... **Asclepiadaceae**

— Anthers free from the style or not, the latter without such appendages.—Leaves above without colleters. Corolla rotate, or campanulate, or funnel-, or salver-shaped, the tube usually longer than
the lobes. .... **Apocynaceae**

1816. Styles or style-branches 2, simple. 1817

— Styles or style brances 3 – more. 1829

1817. Ovary either strictly 1-locular or (in-)completely 2- or 3-locular. 1818

— Ovary (in-)completely 4-locular, or ovaries 4, free.—Ovules 4 *per
flower*... 1828

1818. Stipules present, sometimes early fugacious. 1819

— Stipules absent....................................... 1821

1819. Leaves opposite....................................... 1820

— Leaves alternate.—Woody plants. Flowers 5-merous. Ovary 2-
locular. Ovules 2 per locule, pendulous, anatropous. Fruit a drupe.
Endosperm absent............................ **Dichapetalaceae**

1820. Woody plants. Style 1, bipartite. Ovule 1 per locule. Fruit a drupe.
Rubiaceae

— Herbs. Styles 2, free. Ovules many per locule. Fruit a capsule.
(*Spigeliaceae*: *Mitrasacme*). .... **Loganiaceae**

1821. Micropyle and radicle pointing up or to the centre. Embryo straight.
1822

— Micropyle and radicle pointing down. Embryo curved or plicate.—
Leaves alternate. Ovary either 1-locular and ovules 2–4, or 2-
locular, ovules 1 or 2 per locule.................. **Convolvulaceae**

1822. Ovary 1-locular.—Herbs, rarely undershrubs or shrubs, then flowers
in compound cincinni. 1823

— Ovary 2- or 3-locular. 1826

1823. Ovules 2.—Leaves alternate. Flowers (4- or) 5-merous. 1824

— Ovules many. 1825

1824. Erect plants, rarely twining. Bark without white juice. .. **Icacinaceae**

— Herbaceous climbers with abundant white juice.—Stipules absent.

166

Inflorescence cymose, cincinnoid. Calyx and corolla imbricate. Ovules apical, pendulous. Fruit dry, indehiscent, winged. S.E. Asia to Australia.................................... **Cardiopteridaceae**

1825. Leaves radical or alternate, rarely opposite. Corolla imbricate. Fruit loculicide, rarely septicide and loculicide, or dehiscing irregularily. —Herbs, rarely shrubs or undershrubs, then, as usual, flowers in compound cincinni. .... **Hydrophyllaceae**

— Leaves opposite. Corolla contort, rarely imbricate. Fruit septicide.—Herbs. Style apically slightly bifid. .... **Gentianaceae**

1826. Ovules 2 – more per locule 1827

— Ovule 1 per locule.—Shrublets or woody herbs. Flowers solitary or in dense lateral cincinni. Flowers 4-merous. Ovary 2-locular. Ovule pendulous, anatropous. Africa. (*Wellstediaceae*). ..... **Boraginaceae**

1827. Plants usually herbaceous. Flowers in cincinni. Ovary 2- or 3-locular. Ovules 2 – more per locule. .... **Hydrophyllaceae**

— Ericoid shrubs or undershrubs. Flowers in spikes or capitules. Ovary 2-locular. Ovules 2 per locule.—Leaves alternate, entire. Flowers 5-merous. Ovary slightly immersed in the receptacle. S. Africa.

Bruniaceae

1828. (1817). Micropyle and radicle pointing down. 1993

— Micropyle and radicle pointing up or to the centre.—Leaves undivided, alternate. (incl. *Ehretiaceae*). .... **Boraginaceae**

1829. (1816). Ovary 1, rarely 4, then connate at base. 1830

— Ovaries 3 – 30, free.—Flowers bisexual. Styles 3 – 30. Ovules usually many per ovary. Fruit a capsule..................... **Crassulaceae**

1830. Ovary 1- or 2-locular.—Fruit a capsule. 1831

— Ovary 3 – 16-locular, or ovaries 4, free..................... 1833

1831. Woody plants. Leaves opposite or sub-verticillate. Flowers bisexual. Ovary 2-locular, if 1-locular ovules 3 or 4.—Style 4-fid. Ovules 2 – more. 1832

— Herbs or undershrubs. Leaves alternate. Flowers unisexual. Ovary 1-locular.—Disk present. Ovules 6-more. S. Africa..... **Achariaceae**

1832. Nodes with an interpetiolary ridge or connate stipules. Glandular hairs absent. Ovary 2-locular. Ovules 2 or many per locule. Seeds without an apical tuft of hairs. (*Gelsemieae*).......... **Loganiaceae**

— Stipules absent. Glandular hairs present. Ovary 1-locular. Ovules 3 or 4. Seed with an apical tuft of hairs.—Mexico, C. America. (*Plocospermataceae*)..................................... **Loganiaceae**

1833. Stipules absent. Endosperm present, rarely absent, then style with 4 branches. 1834

— Stipules present, sometimes soon fugacious. Endosperm absent.— Shrubs or trees. Leaves undivided, alternate. Flowers in cymes. Styles or style-branches 3. Ovary 3-locular. Ovules 2 per locule,

pendulous. Fruit a drupe.......................... **Dichapetalaceae**
1834. Flowers unisexual or polygamous.—Woody plants. 1835
— Flowers bisexual, rarely polygamous, then ovules ascending.... 1836
1835. Ovules 1 or 2 per locule.—Leaves undivided. Disk absent. Ovules
pendulous. .... **Ebenaceae**
— Ovules many per locule.—Petals connate at base only. Fruit a
berry. .... **Theaceae**
1836. Styles free, 3 or 5. 1837
— Styles connate at least at base, 2–4. 1838
1837. Herbs. Styles 3.............................. **Hydrophyllaceae**
— Shrub or small tree. Styles 5.—Fruit a drupe. New Caledonia. (also
in *Aquifoliaceae* or *Ebenaceae*)................... **Oncothecaceae**
1838. Style-branches and locules of the ovary or free ovaries 4. Fruit a
drupe or drupelets 4.—Shrubs or trees. (*Ehretiaceae*). **Boraginaceae**
— Style-branches and locules of the ovary (2 or) 3. Fruit a capsule or a
nut:—Plants usually herbaceous................... **Polemoniaceae**
1839. (1681). Ovary 1-locular. 1840
— Ovary 2–more-locular....................................... 1852
1840. Inflorescence not surrounded by a calycoid involucre.......... 1841
— Inflorescence usually surrounded by a calycoid involucre.—Stem
woody. Perianth-segments 4, valvate. Stamens free. Ovules basal, or
apical, or parietal. Endosperm absent................. **Proteaceae**
1841. Ovule 1.. 1842
— Ovules 2–more, sometimes completely immersed in the central
placenta, which then resembles a large, atropous, basal ovule. 1843
1842. Calyx-segments 2 or 5. Corolla-lobes 5, imbricate. Ovule basal.
Fruit a capsule or a nut. 1844
— Calyx 4-dentate. Corolla-lobes 4, valvate. Ovule apical or sub-
parietal. Fruit a drupe.—Endosperm copious. S.E. Asia. (*Cansjera, Lepionurus*)............................... **Opiliaceae**
1843. Calyx-segments 4–7. 1846
— Sepals 2.—Stem herbaceous. Ovules basal. Embryo curved.
Portulacaceae
1844. Calyx-segments 5. Stigmas 5. Embryo straight. 1845
— Calyx-segments 2. Stigmas 1 or 3. Embryo curved.—Herbs.
Basellaceae
1845. Large shrubs. Endosperm absent. (*Aegialitidaceae*). **Plumbaginaceae**
— Herbs, undershrubs, or climbers. Endosperm present. (*Limo-
niaceae*). .... **Plumbaginaceae**
1846. Corolla sometimes with alternipetalous appendages or a confluent
rim, lobes usually imbricate. Disk absent. Ovules ascending.... 1847
— Corolla-lobes usually valvate. Disk present. Ovules pendulous, 2 or
3.—Shrubs or trees. Fruit a drupe.................... **Olacaceae**

1847. Ovules central or basal............................... 1848
— Ovules parietal.—Stipules present. Calyx valvate. Filaments connate. Ovules 2. .... **Sterculiaceae**
1848. Ovules central, if basal immersed in a swollen, central placenta. 1849
— Ovules basal.—Shrubs. Stamens 5, staminodes 5, filiform. Anthers introrse. Ovules 5–7. Fruit a 2-seeded drupe. Arabia to N.W. India. (*Reptonia*)............................... **Sapotaceae**
1849. Anthers dehiscing introrse, or latrorse, or apically. Staminodes rarely present............................... 1850
— Anthers extrorse. Staminodes alternating with the stamens, or discoidally confluent (*Theophrasta*).—Trees or shrubs. Filaments free, rarely connate (*Clavija*). Fruit a berry or a drupe. Tropical America............................... **Theophrastaceae**
1850. Shrubs or trees, rarely herbs or undershrubs. Fruit a berry or a drupe or a viviparous follicle. 1851
— Herbs or undershrubs. Fruit a capsule. Endosperm present.— Flowers bisexual. .... **Primulaceae**
1851. Mangrove treelets. Anthers with a transverse sept. Fruit a viviparous follicle. Endosperm absent. (*Aegicerataceae*).... **Myrsinaceae**
— Plants not from the mangrove. Anthers without transverse septs. Fruit a berry or a drupe. Endosperm present.......... **Myrsinaceae**
1852. (1839). Ovule 1 per locule............................... 1853
— Ovules 2–more per locule.—Ovary 5-locular................ 1855
1853. Style divided. Fruit a capsule or a schizocarp.—Calyx valvate. Corolla imbricate............................... 1854
— Style 1, undivided. Fruit a berry.—Shrubs or trees............ 1856
1854. Herbs. Calyx 5-lobed. Anthers 1-locular. .... **Malvaceae**
— Trees. Calyx 3-lobed. Anthers many-locular.......... **Bombacaceae**
1855. Ovules 3 or more per locule. Style undivided.—Trees. Leaves digitately nerved............................... **Bombacaceae**
— Ovules 2. Style 5-partite.—Calyx valvate. Filaments connate. Disk absent. Ovules ascending......................... **Sterculiaceae**
1856. Leaves undivided. Calyx and corolla imbricate............... 1857
— Leaves usually 3-foliolate, or 1–4 times pinnate, rarely 1-foliolate. Calyx apert. Corolla valvate.—Stipules large, connate with the petiole. Free apical part of the filaments arising outside their tube, arching over it and bearing the anther within. (*Leea*, also included in *Vitaceae*)............................... **Leeaceae**
1857. Calyx and corolla imbricate. 1858
— Calyx apert. Corolla valvate.—Flowers in sessile cauliflorous fascicles. Ovary 5-locular. N. Brazil. (*Brachynema*)........ **Olacaceae**
1858. Flowers in fascicles, rarely solitary, fascicles sometimes on short branchlets. Ovary (2- or 3-) 4–more-locular. Pantropical. **Sapotaceae**

— Flowers in elongated racemes or in panicles. Ovary 2-locular. S.E. Asia. (*Sarcospermataceae*)............................... **Sapotaceae**

1859. (1680). Stamens up to twice as many as the corolla-lobes. 1860
— Stamens more than twice as many as the corolla-lobes......... 1878

1860. Style 1, undivided, stigma 1, undivided or lobed, or sessile..... 1861
— Styles 2–more, free or connate, but rarely as far as the stigmas, or stigmas 2–more, sessile.................................. 1874

1861. Leaves undivided, or digitately or once-pinnately compound. Ovary completely, rarely incompletely 2–more-locular, if 1-locular corolla imbricate, rarely valvate.—Stipules absent. 1862
— Leaves twice-pinnately compound, rarely reduced to the petiole, or absent. Corolla valvate. Ovary 1-locular. (*Mimosoideae*).
Leguminosae

1862. Filaments almost completely connate. Anthers with longitudinal slits. Disk usually present.—Leaves when simple not translucent-glandular-punctate. Ovules 1 or 2 per locule................. 1863
— Filaments free or connate at base, if nearly completely so either leaves translucent-glandular-punctate or anthers with terminal pores. 1864

1863. Leaves usually pinnately compound, rarely simple. Flowers hypogynous. Ovules 1 or 2 per locule. .... **Meliaceae**
— Leaves simple. Flowers epi- or perigynous. Ovule 1 per locule. (*Diclidanthera, Eriandra*)............................. **Polygalaceae**

1864. Leaves undivided, rarely translucent-glandular-punctate, then disk absent and ovules many.................................. 1865
— Leaves undivided, or unifoliolate, or digitately compound, translucent-glandular-punctate. Disk present. Ovules 1 or 2 per locule.
Rutaceae

1865. Disk absent, rarely present but then ovule 1 per locule. 1866
— Disk present.—Leaves usually small and narrow. Stamens 6–10. Anthers usually appendiculate, with 2 more or less apical pores, rarely with 2 longitudinal slits........................ **Ericaceae**

1866. Anthers with 1 pore or transversal slit. 1867
— Anthers with 2 longitudinal slits, if poriform ovules 2 per locule.
1868

1867. Leaves alternate. Anthers with a transversal slit. Ovule 1 per locule. (*Diclidanthera, Eriandra*). .... **Polygalaceae**
— Leaves opposite. Anthers with 1 apical pore. Ovules numerous per locule. .... **Melastomataceae**

1868. Sepals free or connate at base only......................... 1869
— Sepals almost completely connate.—Anthers introrse......... 1872

1869. Floral bracts, if present, not strongly transformed. Corolla imbricate. Ovules 1 or 2 per locule, axillary...................... 1870
— Floral bracts pitcher-like, saccate, or spurred, brightly coloured.

Corolla calyptrate. Ovules many per locule, parietal.—Flowers in spikes, or in racemes, or in umbels. Fruit a tardily dehiscent capsule. Tropical America, West Indies. (*Norantea*).... **Marcgraviaceae**

1870. Latex present. Flowers solitary, or in fascicles, or in racemes, or in panicles. Anthers usually with extrorse slits. Ovule either 1 per locule, or 2 in an incompletely loculed ovary. Fruit a berry or a drupe. 1871

— Latex absent. Flowers in a small dichasial panicle. Anthers dehiscing from the base upward with pore-shaped introrse slits. Ovules 2 per locule, pendulous. Fruit dry, more or less indehiscent. Burma to Yunnan. (*Sladeniaceae*). . . . . **Theaceae**

1871. Flowers in fascicles, rarely solitary, fascicles sometimes on short branchlets. Ovary (1–3-) 4- or more-locular. Pantropical. **Sapotaceae**

— Flowers in elongated racemes or in panicles. Ovary 2-locular (rarely 3-locular) or incompletely 2-locular. S.E. Asia. (*Sarcospermataceae*). . . . . **Sapotaceae**

1872. Ovules 1 or 2 per locule, pendulous. 1873

— Ovules several–many per locule, rarely 1 or 2, then erect and corolla valvate.—Latex absent. Stipules absent. Flowers in simple or compound racemes. Ovary initially 3–5-locular, later incompletely so. . . . . **Styracaceae**

1873. Calyx 5-dentate. Ovary 4- or 5-locular.—Ovule 1 per locule.

Olacaceae

— Calyx 3- or 4-lobed. Ovary 10–18-locular. . . . . **Aquifoliaceae**

1874. (1860). Ovary 1. 1875

— Ovaries 4–30, free, or connate at base only.—Plants usually herbaceous. Flowers bisexual. Ovules many per carpel. . . . **Crassulaceae**

1875. Ovary 2–16-locular. Ovules 1 or 2 per locule. 1876

— Ovary 1- or 5-locular. Ovules many per locule.—Trees. Stipules absent. Flowers unisexual or polygamous, in panicles. Stamens 10. Fruit a berry. . . . . **Caricaceae**

1876. Anthers with 2 slits or pores. Ovules pendulous. 1877

— Anthers with 1 slit. Ovules erect.—Trees. Stipules present. Flowers in terminal panicles. Stamens 10. Filaments connate. Style 2-partite. Ovules 2 per locule. . . . . **Malvaceae**

1877. Stipules present. Flowers in racemes or in panicles.—Flowers unisexual, 5-merous. Ovary 2–4-locular. Ovule 1 per locule.

Euphorbiaceae

— Stipules absent. Flowers solitary or in cymes.—Shrubs or trees. Leaves undivided. . . . . **Ebenaceae**

1878. (1859). Style 1, undivided; stigma 1 or several adjacent at base, or 1, sessile. 1879

— Styles 2–more, free, or more or less completely connate, but not up

171

to the stigmas, or stigmas 2–more, sessile.................. 1888
1879. Ovary 1-locular. 1880
— Ovary 2–more-locular.—Leaves simple, rarely digitately compound.
  1882
1880. Calyx and corolla imbricate. 1881
— Calyx and corolla valvate.—Leaves pinnately compound, or reduced
  to the petiole. Stipules usually present............... **Leguminosae**
1881. Leaves undivided....................................... **Theaceae**
— Leaves pinnately compound.—Trees. Flowers in panicles. Filaments
  connate. Ovary 5-locular. Ovule 1 per locule. .... **Meliaceae**
1882. Ovary 3–25-locular. Ovule 1 per locule.—Woody plants. Anthers
  with 2 longitudinal slits. Embryo straight.................... 1883
— Ovary 2–5(–more)-locular. Ovules 2–more per locule, rarely 1,
  then anthers with 1 slit and embryo curved................... 1884
1883. Flowers solitary, or in glomerules, or in fascicles. Sepals free or
  nearly so, imbricate. Ovary 4–25-locular. Fruit a berry.—Latex
  present... **Sapotaceae**
— Flowers in spikes or racemes. Sepals connate, imbricate or apert.
  Ovary 3-locular.—Stipules absent. .... **Olacaceae**
1884. Stipules absent.—Woody non-resiniferous plants. Leaves undivided.
  Anthers with 2 longitudinal slits............................ 1885
— Stipules present.—Calyx valvate, rarely imbricate, then plants
  resiniferous and calyx enlarged in fruit. Corolla contort........ 1886
1885. Floral bracts pitcher-like, saccate, or spurred, brightly coloured.
  Flowers in terminal racemes, or in spikes, or in umbels.—Calyx imbricate. Corolla calyptrate. Ovules many per locule, parietal. Tropical America, West Indies. (*Norantea*)............. **Marcgraviaceae**
— Floral bracts, if any, not so transformed. Flowers axillary and solitary,
  or in glomerules, or in panicles. (incl. *Sladeniaceae*). .... **Theaceae**
1886. Calyx valvate, rarely apert or closed. Filaments nearly completely
  connate. Anthers with 1 slit, rarely with 2–more, then either ovules
  more than 2 per locule, or ascending. 1887
— Calyx, at least initially, more or less imbricate. Filaments free, or
  connate at base only. Anthers with 2 slits or pores. Ovules 2
  per locule, pendulous or descending.—Woody resiniferous plants.
  Leaves undivided. Flowers in spikes, or in racemes, or in panicles.
  Calyx usually enlarged in fruit. Ovary 3-locular. .... **Dipterocarpaceae**
1887. Filaments connate into 1 bundle.—Leaves simple. Anthers with 1
  slit. Pollen spinose. .... **Malvaceae**
— Filaments free or usually connate into 2–more bundles.—Woody
  plants. .... **Bombacaceae**
1888. (1878). Ovary 2–more-locular, rarely 1-locular, then either ovaries
  2–more, free, or ovule 1..................................... 1889

172

— Ovary 1, 1-locular. Ovules many.—Trees. Leaves undivided. Stipules absent. Flowers in panicles. Calyx-lobes 3–5, valvate. Corolla-lobes 11–14, imbricate. Style 2-partite. Tropical W. Africa.

Hoplestigmataceae

1889. Ovary 1, 2–more-locular, rarely 1-locular, then ovule 1. Corolla-lobes 3–8, rarely more, then calyx imbricate. 1890
— Ovaries 2–several, free, 1-locular. Ovules several per ovary. Calyx and corolla (4- or) 5-lobed, valvate.—Woody plants. Leaves alternate, pinnately compound. Stipules present. (*Affonsea, Archidendron*). . . . . **Leguminosae**

1890. Anthers with 2 slits or pores.—Woody plants. 1891
— Anthers with 1 slit.—Stipules present. Flowers 5-merous, bisexual or polygamous. Calyx valvate. Corolla contort. Filaments connate.

Malvaceae

1891. Calyx imbricate, rarely valvate, then stipules absent. 1892
— Calyx valvate.—Stipules present. Flowers bisexual. Calyx 3-lobed. Corolla contort, 5-partite. Filaments connate. Ovary 2-locular. Ovule 1 per locule, ascending. (? *Scleronema* from tropical S. America). . . . . **Bombacaceae**

1892. Leaves undivided. 1893
— Leaves digitately divided or compound.—Flowers in racemes, bisexual. Calyx and corolla deeply partite, imbricate. Ovary 4–6-locular. Ovule 1 per locule, ascending. Endosperm absent or nearly so. Tropical America. . . . . **Caryocaraceae**

1893. Leaves opposite or in whorls. 1894
— Leaves usually alternate.—Ovule 1 per locule, pendulous, or 2–more. 1895

1894. Stipules present. Flowers in racemes or in panicles. Ovules ascending. Endosperm absent. Tropical S. America. . . . . **Quiinaceae**
— Stipules absent. Flowers solitary or in cymes. Ovules pendulous. Endosperm present. . . . . **Ebenaceae**

1895. Ovules 1 or 2 per locule. Endosperm copious.—Flowers unisexual or polygamous, solitary or in cymes. . . . . **Ebenaceae**
— Ovules 2–more per locule. Endosperm scanty or absent.—Calyx and corolla deeply divided, imbricate. **Theaceae**

1896. (1573). Fertile stamens less than the corolla-lobes, 1–4, rarely as many, then 2. 1897
— Fertile stamens either as many as the corolla-lobes and more than 2, or more. 1944

1897. Ovary 1, 1-locular or nearly so. 1898
— Ovary 1, 2–more-locular or nearly so, or ovaries 4 or 5, free, or connate at base only. 1917

1898. Ovule 1. 1899

173

— Ovules 2–more. 1904
1899. Flowers bisexual. 1900
— Flowers unisexual.—Male flowers with a 2–4-lobed corolla and 2 or 3 stamens. Female flowers with an undivided or 2-lobed corolla and 3 stigmas. . . . . **Menispermaceae**
1900. Stamens 4. 1901
— Stamens 1 or 2.—Leaves radical. Flowers polygamous, in a spike-like capitule. Corolla 3–5-dentate. Stigmas 1 or 2. . . . **Plantaginaceae**
1901. Leaves alternate. Endosperm fleshy. Embryo straight. 1902
— Leaves opposite. Endosperm absent. Embryo plicate.—Anthers with 2 longitudinal slits. 1903
1902. Flowers in capitules, rarely in spikes. Anther with 1 transversal slit. Stigma capitate or 2-lobed. Fruit dry, indehiscent.—Ovule pendulous. (incl. *Poskea*, sometimes included in *Ehretiaceae, Boraginaceae*). . . . . **Globulariaceae**
— Flowers in spikes. Anthers with longitudinal slits. Stigma undivided, not thickened. Fruit usually a capsule. . . . . **Scrophulariaceae**
1903. Herbs. Flowers in spikes. Stigma 2-lobed. Ovule sub-basal, atropous. Fruit a nut. Temperate E. Asia and N. America. . . . . . **Phrymaceae**
— Shrubs or climbing undershrubs. Flowers solitary or in fascicles. Stigma 2-partite. Ovule anatropous. Fruit a drupe. Tropical Africa and America. (*Mendonciaceae*). . . . . **Acanthaceae**
1904. Ovules 2–4. 1905
— Ovules 8–more. 1911
1905. Terrestrial prostrate or erect herbs, or shrubs, or climbers. 1906
— Aquatics.—Flowers solitary. Stamens 2. Ovules 2. (*Utricularia*).
Lentibulariaceae
1906. Woody plants or prostrate herbs. Stamens 4. 1907
— Erect herbs. Stamens 2 or 3. 1910
1907. Ovules 2. 1908
— Ovules 3 or 4.—Leaves opposite or in whorls. Ovary incompletely locular. (incl. *Symphoremataceae: Congea*). . . . . **Verbenaceae**
1908. Shrubs or climbers. Stamens inserted above the base of the corolla-tube. 1909
— Prostrate herbs. Stamens adnate to the base of the corolla-tube.—Flowers solitary. Calyx deeply divided. W. equatorial and S.W. tropical Africa. (*Linariopsis*). . . . . **Pedaliaceae**
1909. Flowers in racemes. Calyx deeply lobed. Stamens adnate to the middle of the corolla-tube.—Shrubs. . . . . **Verbenaceae**
— Flowers solitary or in fascicles. Calyx slightly lobed. Stamens adnate to the upper part of the corolla-tube.—Shrubs or climbers. Fruit a drupe. (*Mendonciaceae*). . . . . **Acanthaceae**

174

1910. Flowers in racemes. Stamens 2. Ovules 4. Mexico. (*Martynia*).
   Martyniaceae
— Flowers in cymes. Stamens 3. Ovules 3. . . . . **Portulacaceae**
1911. Fertile stamens 2 or 4 and either corolla-lobes 5, or staminodes not
   well-developed. 1912
— Fertile stamen 1, rarely 2, then with 2 smaller staminodes. Corolla-
   lobes 4. . . . . **Gentianaceae**
1912. Placenta 1, central. 1913
— Placentas 2 – 4, parietal. 1914
1913. Calyx deeply divided. Stamens 2, adnate to the base of the corolla-
   tube. Endosperm absent.—Herbs. Leaves radical or alternate.
   Anthers with 1 transversal slit. . . . . **Lentibulariaceae**
— Calyx shortly lobed. Stamens 2 or 4, inserted on the corolla-tube.
   Endosperm present. . . . . **Scrophulariaceae**
1914. Plants not parasitic. Leaves well-developed, green. 1915
— Parasitic herbs. Leaves scale-like.—Flowers solitary, terminal, or in
   spikes, or in racemes. Stamens 4. Fruit a capsule. . . **Orobanchaceae**
1915. Fruit a capsule, or a nut, or a berry, endocarp not indurated. . . 1916
— Fruit a horned 4-locular capsule, endocarp indurated.—Erect or
   prostrate herbs. Leaves simple. Flowers in racemes. Corolla-lobes 5,
   short, slightly unequal, imbricate. Stamens 4, inserted on the corolla-tube. Pollen large, reticulate, without pores. Disk regular.
   Placentas 2-partite. Stigma 2-partite. Tropical and subtropical
   America. . . . . **Martyniaceae**
1916. Leaves usually pinnately compound, rarely simple. Corolla-lobes descendingly imbricate. Seeds rather large, flat, usually winged or with
   a prominent margin, immersed in the enlarged, usually fleshy
   placentas.—Woody plants. Stamens 4. Disk present. Stigma 2-
   partite. Fruit usually an elongated berry, or dry and indehiscent, or
   a capsule. Endosperm absent. . . . . **Bignoniaceae**
— Leaves simple, undivided. Corolla-lobes usually ascendingly imbricate. Seeds small, not immersed in the placentas. . . . . . . **Gesneriaceae**
1917. (1897). Ovary 2-, rarely 3-locular. 1918
— Ovary 4 – 10-locular, or ovaries 4 or 5, free. 1933
1918. Stipules absent, nodes rarely with interpetiolary lines.—Leaves
   opposite or alternate. Stamens 2 – 4. 1919
— Stipules present or nodes with thin interpetiolary lines.—Woody
   plants. 1920
1919. Corolla imbricate, not plicate, rarely valvate or plicate, then leaves
   opposite. Sept of the ovary usually transverse to the plane of symmetry of the flower. 1921
— Corolla valvate or plicate, then sometimes also imbricate. Sept of
   the ovary usually oblique to the plane of symmetry of the flower.—

175

Leaves alternate, sometimes in pairs, but not opposite. Flowers solitary or in cymes. Ovules several – many per locule. Fruit a septicide capsule or a berry. (incl. *Salpiglossidaceae*)............. **Solanaceae**

1920. Leaves opposite. Nodes with thin interpetiolary lines. Stamen 1. Ovules many per locule.—Corolla-lobes 4, valvate. Tropical W. Africa. (*Antoniaceae: Usteria*). .... **Loganiaceae**

— Leaves alternate. Stipules present, often early fugacious. Stamens 2 or 3. Ovules 2 per locule....................... **Dichapetalaceae**

1921. Leaves usually alternate. Ovule 1 per locule, pendulous or descending, or 2 – more, rarely 1 and erect or ascending, or 2 and separated by a sept, then fruit a loculicide capsule with hook-shaped funicles or with the micropyle and radicle pointing upwards.—Thecae usually confluent. 1924

— Leaves usually opposite or in whorls. Ovule 1 per locule, erect or ascending, or ovary incompletely locular and ovules 2. Micropyle and radicle pointing downwards. Fruit a drupe, or a schizocarp, or a septicide capsule. Seeds sessile.—Thecae usually separate..... 1922

1922. Endosperm present...................................... 1923

— Endosperm absent.—Not with the combination of characters of next lead... **Verbenaceae**

1923. Ericoid undershrubs. Leaves in whorls. Flowers in racemose spikes. Anthers inappendiculate. Ovule basal.—S. Africa. (*Stilbaceae*).

Verbenaceae

— Herbs. Leaves opposite. Flowers solitary, axillary, or in few-flowered cymes. At least some anthers appendiculate at base. Ovule axillary. (*Dicrastylidaceae-Achariteae*). .... **Verbenaceae**

1924. Fruit a capsule or a berry, rarely dry, indehiscent, or a drupe, then *either* calyx undivided, *or* flowers in capitules, or in spikes, or in panicles, *or* thecae separate and disk well-developed. 1925

— Fruit a drupe or a nut.—Shrubs or trees, rarely undershrubs. Leaves usually alternate, undivided or lobed. Flowers solitary or in fascicles. Calyx 5-partite. Disk absent or indistinct. Stamens 4. Thecae confluent at the apex. Ovules 1 – 8 per locule, pendulous, anatropous. Seeds few. Endosperm scanty. .... **Myoporaceae**

1925. Endosperm usually copious, rarely absent or scanty, then stigma undivided and fruit a septicide or both septicide and loculicide capsule, or dry and indehiscent. Cotyledons usually narrow.—Leaves simple, sometimes deeply incised. Fruit a schizocarp, or dry and indehiscent, or a berry, or a capsule, then when loculicide *either* sepals connate up to halfway or more, *or* corolla nearly actinomorphic and 4-fid, *or* anthers with 1 slit, *or* stigma simple. Seeds usually minute.

1926

— Endosperm very scanty and almost membranous, or absent, rarely well-developed (*Acanthaceae*) but then copious and sepals connate at base only, corolla bilabiate or nearly actinomorphic, 5-lobed, anthers with 2 longitudinal slits or pores, stigma 2–4-lobed, and fruit a loculicide capsule. Cotyledons usually broad.—Stigma lobed or partite, rarely simple, then fruit either a loculicide or irregularily dehiscing capsule, or a berry, or a drupe. 1927

1926. Ericoid herbs or undershrubs. At least the lower leaves opposite. Corolla 4- or 5-lobed, upper lobes covering the 2 lateral or basal ones in bud. Stamens 2 or 4, if 5 the dorsal staminodial. Ovule 1 per locule, apical, pendulous. Fruit a 1-seeded drupe, or a schizocarp with 2 nutlets. S. Africa, Madagascar. (*Selaginaceae*).
Scrophulariaceae

— Plants otherwise. . . . . **Scrophulariaceae**

1927. Fruit not a schizocarp. 1928

— Fruit a schizocarp of 4 nutlets.—Herbs. Leaves alternate. Style gynobasic. Ovule 1 per locule, epitropous, basal, erect. Endosperm absent. Radicle pointing upwards. . . . . **Boraginaceae**

1928. Endosperm absent, rarely scanty, then fruit a capsule without wings or spines and disk indistinct. 1929

— Endosperm scanty, almost membranous.—Plants usually herbaceous, with capitate glandular hairs. Leaves dentate or deeply incised. Flowers solitary or in fascicles. Stamens 4. Disk distinct. Stigma partite. Fruit a winged or spiny nut or a capsule. Embryo straight. . . . . **Pedaliaceae**

1929. Plants usually herbaceous. Leaves simple, incised or not. Calyx usually deeply incised, or sepals free. 1930

— Plants usually woody. Leaves usually compound. Sepals nearly completely connate, rarely at base only.—Calyx apert, or closed, or valvate. Stigma 2-partite. Fruit a more or less juicy berry, or a septicide or loculicide capsule. Placentas in fruit usually separated by an elongated sept. Seeds several–many, laterally attached, sessile or nearly so, winged, rarely not, then either fruit a berry or leaves compound and seeds in 1 row. Endosperm absent. . . . . . **Bignoniaceae**

1930. Nodes usually swollen. Leaves usually with cystoliths. Fruits usually with indurated, hook-shaped, rarely wart-shaped funicles, or sessile, then either ovules 1 or 2 per locule, or endosperm present, or sepals connate at base only.—Leaves simple, sometimes partite. Fruit a loculicide capsule, rarely a drupe. Placentas in fruit approximate to fused. Seeds not winged, usually 2–10 per locule in 2 rows, rarely solitary. 1931

— Nodes usually not swollen. Leaves without cystoliths. Funicles not indurated.—Herbs, sometimes woody at base. Leaves undivided. Calyx imbricate, 5-partite. Stigma undivided. Fruit irregularily dehiscent or a berry. .... **Gesneriaceae**

1931. Ovary 2-locular. Fruit a 2-many-seeded capsule. 1932

— Ovary 1- (or 2-)locular. Fruit a 1- or 2-seeded drupe. (*Mendonciaceae*). .... **Acanthaceae**

1932. Ovules 1–many per locule. Hardened funicles well-developed.
Acanthaceae

— Ovules 2 per locule. Hardened funicles absent to papillate. (*Thunbergiaceae*). .... **Acanthaceae**

1933. (1917). Ovule 1 per locule. 1934

— Ovules 2–more per locule. 1940

1934. Leaves alternate, rarely opposite, then ovules pendulous and micropyle and radicle directed upwards. 1935

— Leaves opposite or in whorls, exceptionally alternate.—Ovules either basal, or micropyle and radicle directed downwards. 1936

1935. Flowers in cymes, or in racemes, or in panicles. Fertile stamen 1. Ovary deeply 4-partite. Fruit dry, indehiscent. .... **Boraginaceae**

— Flowers solitary or in fascicles. Fertile stamens 4. Ovary undivided or nearly so. Fruit a drupe or a nut.—Shrubs or trees, rarely undershrubs. Leaves alternate, rarely opposite. Corolla-lobes 5. Anthers with 1 slit. Disk indistinct or absent. Stigma 1. Ovules pendulous, micropyle and radicle directed upwards. Endosperm scanty.
Myoporaceae

1936. Fertile stamens 2 or 4.—Micropyle and radicle directed downwards.
1937

— Fertile stamens 4.—Flowers solitary. Ovary undivided, 4–8-locular. Ovules basal. Fruit spinose, dry, indehiscent. Endosperm scanty.
Pedaliaceae

1937. Ovary undivided or nearly so, rarely distinctly lobed, then initially incompletely locular, ovules inserted in the middle and mericarps more or less drupaceous. Ovules pendulous or inserted in the middle, rarely basal, then flowers in spikes, or in racemes, or in capitules. 1938

— Ovary deeply 4-partite, usually to the base, rarely less, completely 4-locular, then, as usual, mericarps dry, rarely drupaceous. Ovules basal, rarely inserted somewhat higher or halfway.—Flowers usually in false whorls. .... **Labiatae**

1938. Ovule either axillary, campylotropous, or basal, anatropous. 1939

— Ovule apical, atropous.—Climbing shrubs. Flowers in involucrate capitules. Endosperm absent. S.E. Asia. (*Symphoremataceae: Congea*). .... **Verbenaceae**

1939. Herbs, undershrubs, or shrubs. Flowers 1–3 together, in axillary cymes. Flowers ± bilabiate. Anthers inappendiculate. Ovule axillary, campylotropous. Fruit a drupe or a schizocarp. Endosperm

178

present. Australia. (*Dicrastylidaceae-Chloantheae*). .... **Verbenaceae**

— These characters not combined. .... **Verbenaceae**

1940. Leaves simple, translucent-glandular-punctate or not. Stamens 4. Ovary undivided or obscurely lobed. 1941

— Leaves usually compound, translucent-glandular-punctate. Stamens 2 or 3. Ovary deeply divided.—Ovules 2 per locule. ...... **Rutaceae**

1941. Leaves entire, or lobed, or incised. Tendrils absent. 1942

— Leaves deeply incised. Tendrils present.—Herbs. Flowers in racemes. Anthers with 2 slits. Disk saucer-shaped. Stigma 4-lobed. Ovules 3 per locule. Fruit a spiny capsule. Endosperm absent. C. America to Peru. (*Tourrettia*). .... **Bignoniaceae**

1942. Leaves not translucent-glandular-punctate. Anthers with 2 slits. 1943

— Leaves translucent-glandular-punctate. Anthers with 1 slit or 2 apically confluent ones.—Shrubs or trees, rarely undershrubs. Leaves undivided. Flowers solitary or in fascicles. Disk indistinct or absent. Stigma 1, undivided or lobed. Fruit a drupe or a nut. Endosperm scanty..................................... **Myoporaceae**

1943. Stigma undivided or 2-lobed. Ovary incompletely locular.—Herbs. Leaves undivided. Disk distinct. Ovules numerous. Fruit a capsule or a berry. Endosperm absent. .... **Gesneriaceae**

— Stigmas 2–4, or stigma 1, 2–4-partite. Ovary 2–4-locular.—Herbs or undershrubs, rarely shrubs, with glandular hairs. Leaves undivided, or lobed, or divided. Disk present. Fruit a capsule or a nut. Endosperm scanty................................. **Pedaliaceae**

1944. (1896). Fertile stamens as many as the corolla-lobes, rarely more, then stamens 3 or 4.. 1945

— Fertile stamens more than the corolla-lobes, 5–more.......... 1994

1945. Ovary apically completely closed.......................... 1946

— Ovary apically open.—Herbs. Petals 2, fimbriate. Stamens 3, excentric. Ovary 1-locular. Stigmas 4, sessile. Ovules numerous. (*Oligomeris, Resedella*). .... **Resedaceae**

1946. Stamens as many as the corolla-lobes, epipetalous. 1947

— Stamens as many as the corolla-lobes, alternipetalous, or more, then stamens 3 or 4. 1950

1947. Flowers usually bisexual..................................... 1948

— Flowers unisexual.—Woody plants. Male flowers with a 2–8-lobed corolla and connate filaments. Female flowers with 1 or 2 petals or 2 corolla-lobes and 3 stigmas. Ovule 1, pendulous. Fruit a drupe. Endosperm scanty. .... **Menispermaceae**

1948. Inflorescence-axis without a calycoid involucre. Corolla imbricate. Fruit a capsule.—Herbs or undershrubs....................... 1949

— Inflorescence-axis often with a calycoid involucre. Perianth-segments

4, valvate. Fruit a nut or a drupe, or a follicle, or a capsule.—Plant usually woody. Stigma 1. Endosperm absent........... **Proteaceae**

1949. Stigma 1. Ovules 2–more.—Small, ericoid undershrubs. Mediterranean, N.E. Africa. (*Coridaceae*)..................... **Primulaceae**

— Stigmas 5. Ovule 1.—Sepals with long glandular hairs. (*Plumbago*).
Plumbaginaceae

1950. Ovary 1-locular or nearly so. 1951

— Ovary completely 2–more-locular or nearly so, or ovaries 2–more, free... 1962

1951. Ovule 1. 1952

— Ovules 2–more.—Corolla-lobes 3–8, imbricate. 1954

1952. Corolla-lobes 3 or 4, imbricate. Stamens 4. Stigmas 1 or 2, without a cupular involucre. N. temperate zone..................... 1953

— Corolla-lobes 5, valvate. Stamens 5. Stigma 1, surrounded by a cupular involucre.—Herbs. Leaves radical. Flowers in capitules. Ovule basal. Anthers with 2 longitudinal slits. Australia.
Brunoniaceae

1953. Leaves opposite. Flowers in spikes. Anthers with 2 longitudinal slits. Stigmas 2. Ovule erect, atropous..................... **Phrymaceae**

— Leaves alternate. Flowers in capitules. Anthers with 1 transversal slit. Ovule pendulous, anatropous.................. **Globulariaceae**

1954. Ovules 5–more... 1955

— Ovules 2–4... 1956

1955. Leaves individed. Style 1, stigmas 1 or 2..................... 1958

— Leaves usually divided. Style 2-fid.—Herbs. Corolla nearly actinomorphic. Stamens 5......................... **Hydrophyllaceae**

1956. Anthers with longitudinal slits. 1957

— Anthers with 1 terminal pore.—Perennial herbs or shrubs. Leaves usually densely pubescent. Sepals 4 or 5, unequal, free, imbricate. Ovules 2, collateral, parietal, pendulous. Fruit indehiscent with bristles or spines. Endosperm absent. America.......... **Krameriaceae**

1957. Leaves opposite or in whorls. Ovary incompletely 1-locular.
Verbenaceae

— Leaves alternate. Ovary completely 1-locular.—Flowers in fascicles. Cuba. (*Goetzeaceae*: *Henoonia*)..................... **Solanaceae**

1958. Plants autotrophic. Leaves green......................... 1959

— Plants parasitic, non-green.—Herbs. Leaves scale-like. Flowers solitary, or in spikes, or in racemes. Stamens 4. Endosperm copious. Embryo indistinct. .... **Orobanchaceae**

1959. Woody plants. Stipules or a stipular sheath present. 1960

— Plants usually herbaceous. Stipules absent................... 1961

1960. Leaves alternate. Flowers in spikes, or in racemes. Stamens 5, connate at base. Fruit a capsule......................... **Violaceae**

180

— Leaves opposite. Flowers in cymes. Stamens 5–8(–16), adnate to the corolla-tube. Fruit a berry. (*Potaliaceae*). .... **Loganiaceae**

1961. Leaves opposite, entire. Corolla nearly actinomorphic, usually contort. Stigma 2-lobed. Endosperm copious. .... **Gentianaceae**

— Leaves various. Corolla usually zygomorphic, often bilabiate, imbricate. Style 1, stigma capitate or 2-lobed. Endosperm scanty or absent. .... **Gesneriaceae**

1962. (1950). Ovary 2- or 3-locular, or ovaries 2, free at base but connate by the styles. 1963

— Ovary 4–20-locular, or ovaries 4 or 5, free. 1967

1963. Ovary 2- or 3-locular. 1964

— Ovaries 2, free at base but connate by the styles.—Herbs. Leaves opposite. Corolla valvate. Stamens connate into a ring and adnate to the style-apex. Pollen united into pollinia. (*Ceropegia*).

Asclepiadaceae

1964. Ovary 3-locular. 1965

— Ovary 2-locular. 1968

1965. Anthers with 2 longitudinal slits. Stigmas 3. 1966

— Anthers with 1 pore. Stigma 1.—Woody plants. Stipules absent.

Polygalaceae

1966. Woody plants. Stipules present, often minute. ...... **Dichapetalaceae**

— Herbs. Stipules absent. .... **Polemoniaceae**

1967. Leaves simple or ovary divided. Embryo usually straight. 1989

— Leaves compound, 1–7-foliolate, translucent-glandular-punctate. Ovary 2–4-partite. Ovules 2 per locule. Embryo usually curved.

Rutaceae

1968. Leaves alternate, at least the upper, sometimes in pairs but not opposite, or all radical. 1969

— Leaves opposite or in whorls. 1975

1969. Style undivided, if 2-partite plants woody and endosperm absent.

1970

— Style 2-partite.—Herbs. Leaves usually undivided. Flowers 5-merous. Endosperm present. .... **Hydrophyllaceae**

1970. Ovules 2 per locule.—Woody plants. Leaves undivided. 1971

— Ovules 4–more, rarely 1 per locule. 1973

1971. Stipules absent. Flowers solitary or in terminal few-flowered racemes.—Corolla 5-lobed. 1972

— Stipules present, often inconspicuous. Flowers in fascicles.—Stigmas 2. .... **Dichapetalaceae**

1972. Plants usually climbing with tendrils. Flowers solitary. Corolla plicate. Stigma 1. .... **Convolvulaceae**

— Erect woody plants. Flowers solitary or in terminal few-flowered racemes. Corolla valvate. Stigmas 2.—Cuba. (*Goetzeaceae: Espa-*

daea). .... **Solanaceae**
1973. Stigmas 2. 1974
— Stigma 1. 1976
1974. Leaves simple. Endosperm present. .... **Scrophulariaceae**
— Leaves 1–3-foliolate. Endosperm absent.—Woody plants. Filaments free. Seeds winged. .... **Bignoniaceae**
1975. Stamens adnate to the corolla. 1982
— Stamens free from the corolla.—Shrublets. Leaves in whorls. Stamens 4. .... **Ericaceae**
1976. All anthers or filaments connate or nearly so. 1977
— Anthers all free, ·or connate in pairs and stamens adnate to the corolla. 1978
1977. Anthers 5, all connate, with 2 introrse slits. Ovules numerous.—Corolla usually valvate. Embryo straight. (*Lobeliaceae*).
Campanulaceae
— Filaments usually completely connate, anthers free, erect, with 1 pore. Ovule 1 per locule.—Stigma 1. .... **Polygalaceae**
1978. Ovule 1 per locule. 1979
— Ovules numerous per locule. 1980
1979. Filaments free from the corolla. Endosperm absent.—Leaves undivided or lobed. Anthers with 2 longitudinal slits. Stigma 1. .. **Verbenaceae**
— Filaments adnate to the corolla. Endosperm present.
Scrophulariaceae
1980. Corolla imbricate, not plicate. Sept of the ovary at a right angle to the plane of symmetry of the flower. Fruit dehiscing longitudinally or with pores, rarely indehiscent, then seeds 1 or 2. Embryo straight or slightly curved. 1981
— Corolla usually valvate or plicate, then sometimes also imbricate, rarely imbricate but not plicate, then fruit with a lid or indehiscent and seeds many and embryo usually strongly curved. Sept of the ovary usually oblique to the plane of symmetry of the flower.—Endosperm present. .... **Solanaceae**
1981. Seeds winged. Endosperm absent.—Shrubs. Leaves usually 1- or 3-foliolate. Fruit a capsule with 2 valves. Tropical Africa to S. Africa, Madagascar. (*Catophractes, Rhigozum*). .... **Bignoniaceae**
— Seeds not winged. Endosperm present.—Fruit dehiscing longitudinally or with pores, rarely indehiscent. .... **Scrophulariaceae**
1982. Stipules or a stipular sheath absent. 1983
— Stipules or a stipular sheath present.—Leaves undivided or lobed. Flowers nearly actinomorphic. Ovules numerous per locule. Endosperm present. .... **Loganiaceae**
1983. Corolla imbricate. 1984
— Corolla valvate.—Leaves in whorls. Corolla 5-lobed. Filaments and

anthers connate. . . . . **Campanulaceae**

1984. Seeds sessile or on a short hardly indurated funicle. 1985

— Seeds on elongated, indurated, more or less hook-shaped funicles.—
Fruit a loculicid capsule. Endosperm absent. (*Acantheae*).

Acanthaceae

1985. Ovules either 1 per locule, or 2 and collateral. Micropyle and radicle directed downwards. 1986

— Ovules either 1 per locule, or 2 and serial, or more. Micropyle and
radicle directed upwards.—Endosperm copious. 1988

1986. Flowers in capitules with an involucre of 5, or 6, or 2 deeply 3-lobed
bracts.—Lianas. Endosperm absent. S.E. Asia. (*Symphoremataceae*:
Sphenodesme). . . . . **Verbenaceae**

— Flowers rarely in capitules, then without such an involucre. 1987

1987. Erect shrubs. Flowers in axillary, spike-like cincinni. Endosperm
present.—Madagascar. (*Dicrastylidaceae*: *Acharitea*). . . **Verbenaceae**

— Plants otherwise. Endosperm absent. . . . . **Verbenaceae**

1988. Style undivided. . . . . **Scrophulariaceae**

— Style 2-partite.—Flowers 5-merous, nearly actinomorphic.

Hydrophyllaceae

1989. (1967). Ovary 4-locular, or ovaries 4, free. 1990

— Ovary 5–20-locular.—Woody plants. Leaves alternate, undivided.
Stamens 5–10, free from the corolla. Anthers with apical pores.
Ovules many per locule. . . . . **Ericaceae**

1990. Ovules many per locule. 1991

— Ovule 1 per locule. 1992

1991. Leaves alternate, undivided or lobed. Tendrils absent. Flowers solitary. Corolla nearly actinomorphic. Stamens 5. Seeds not winged.
Endosperm present. . . . . **Solanaceae**

— Leaves opposite, deeply incised. Tendrils present. Flowers in spikes.
Corolla bilabiate. Stamens 4. Seeds usually winged. Endosperm ab-
sent. . . . . **Bignoniaceae**

1992. Leaves opposite or in whorls, very rarely alternate, then ovary undivided. Ovules atropous or apotropous, micropyle and radicle
pointing downwards. 1993

— Leaves alternate. Ovules epitropous, micropyle and radicle pointing
upwards or to the axis, rarely downwards.—Leaves undivided,
usually hispid. Flowers usually in secund cincinni. Ovary deeply par-
tite. . . . . **Boraginaceae**

1993. Flowers usually in false whorls. Ovary deeply divided, usually to the
base, rarely less, but still distinctly lobed, completely locular and
then, as usual, mericarps dry. Ovule basal, rarely inserted somewhat
higher up or in the middle. . . . . **Labiatae**

— Flowers in spikes, or in racemes, or in capitules. Ovary undivided or nearly so, rarely distinctly lobed, then initially incompletely locular. Ovule pendulous or laterally attached, inserted in the middle or above, rarely basal. Mericarps drupaceous. .... **Verbenaceae**

1994. (1944). Ovary 1-locular. 1995

— Ovary 2–20-locular.—Leaves usually undivided. 1999

1995. Leaves usually undivided. Stipules absent or nodes with an annular gland. 1996

— Leaves usually compound. Stipules present.—Stamens 10. Filaments free. Anthers with 2 longitudinal slits. Ovules 2–8. .... **Leguminosae**

1996. Bark inside with tough silky fibres. Corolla slightly developed, more or less annular.—Ovule 1. .... **Thymelaeaceae**

— Bark inside without such fibres. Corolla well-developed. 1997

1997. Stamens free or connate, not on an androgynophore, all fertile. 1998

— Stamens on an androgynophore, 4 fertile and 4 or 5 staminodial.— Ovule 1, basal. Australia. (*Emblingiaceae*). .... **Capparaceae**

1998. Woody plants. Leaves simple. Sepals and corolla-lobes 5. Stamens usually 8, never 6, at base adnate to the corolla. Tropics (*Xantho-phyllaceae*). .... **Polygalaceae**

— Herbs. Leaves compound. Sepals 2. Corolla-lobes 4. Stamens 6, free from the corolla. Temperate regions. (*Fumariaceae*).   **Papaveraceae**

1999. Ovule 1 per locule.. 2000

— Ovules 2–more per locule.—Stem woody.................... 2002

2000. Filaments connate.—Stamens 7 or 8. Anthers with 1 pore or with longitudinal slits. Style 1. .... **Polygalaceae**

— Filaments free.—Leaves undivided. Stamens 6–10. Anthers with 2 longitudinal slits.. 2001

2001. Bark of twigs inside with tough silky fibres. Stem woody. Style 1.— Stipules absent. Stamens 8–10................... **Thymelaeaceae**

— Bark of twigs without such fibres. Herbs. Styles 3.—Flowers unisexual. Stamens 6–10........................... **Euphorbiaceae**

2002. Leaves usually digitately compound. Stipules present.—Flowers 5-merous.. **Bombacaceae**

— Leaves simple. Stipules absent.—Stamens 6–18. Anthers with 2 apical pores, exceptionally with 2 longitudinal slits. Ovules 2–more per locule. .... **Ericaceae**

2003. (1572). Fertile stamens less than the corolla-lobes, 1–4........ 2004

— Fertile stamens as many as the corolla-lobes or more (some *Lecy-thidaceae*, e.g. *Asteranthaceae*, *Napoleonaeaceae* have a 20–40-rayed corolla (? = connate staminodes) and 10–many fertile stamens). 2018

2004. Ovules 2 or more *per ovary*. 2005

— Ovules 1 *per ovary*, sometimes also a few abortive ones present, or some locules empty.. 2009

2005. Ovary 1. 2006
— Ovaries 2–more, free.—Ovules usually numerous. 2014
2006. Ovules 2 *per ovary.* 2007
— Ovules numerous *per ovary.*—Leaves opposite. Flowers bisexual.
Corolla actinomorphic, valvate or slightly imbricate. Stamens 2,
adnate to the corolla. Anthers straight, introrse or latrorse. Disk
present. Stigmas 1 or 2. Ovary completely 2-locular. . . . . . **Rubiaceae**
2007. Leaves opposite. Flowers bisexual. Corolla imbricate. Anthers introrse.—E. Asia. 2008
— Leaves alternate. Flowers unisexual. Corolla valvate. Anthers ex-
trorse.—Climbing or prostrate herbs or undershrubs. Stamens 2 or
3. Ovary 1-locular. . . . . **Cucurbitaceae**
2008. Aquatic herbs. Stamens 2. Staminodes 2. Ovary with 1 fertile and 1
sterile locule. (*Trapellaceae*). . . . . **Pedaliaceae**
— Shrubs. Stamens 4. Ovary with 2 fertile and 2 sterile locules.
(*Dipelta*). . . . . **Caprifoliaceae**
2009. Ovary 1-locular, or 3-locular with 1 fertile and 2 empty locules. 2010
— Ovary 3-locular with 1 locule with 1 fertile ovule and 2 with several
abortive ones.—Shrubs. Leaves opposite or in whorls, undivided.
Stipules absent. Flowers bisexual, solitary or in cymes. Corolla
slightly zygomorphic, imbricate. Stamens 4. Anthers introrse. Stigma 1. Endosperm present. (*Linnaea*). . . . . **Caprifoliaceae**
2010. Leaves opposite, or in whorls, or all radical. Flowers bisexual or
polygamous. Corolla imbricate. Anthers introrse.—Stipules absent.
Fruit dry, indehiscent. 2011
— Leaves alternate. Flowers unisexual. Corolla valvate. Anthers extrorse.—Ovary 1-locular. . . . . **Cucurbitaceae**
2011. Flowers in capitules, rarely in axillary whorls, or in dichasia. Epicalyx present. Ovary 1-locular. 2012
— Flowers in cymes or in dichasia. Epicalyx absent. Ovary 3-locular,
with 1 fertile and 2 sterile locules. . . . . **Valerianaceae**
2012. Flowers not in dichasia. Epicalyx simple. 2013
— Flowers in dichasia. Epicalyx double.—Inflorescence glandular. S.E.
Asia to New Guinea. (*Triplostegiaceae*, also in *Valerianaceae*).
                                                                    Dipsacaceae
2013. Flowers in axillary whorls. (*Morinaceae*). . . . . **Dipsacaceae**
— Flowers in capitules. . . . . **Dipsacaceae**
2014. Flowers nearly always bisexual. Corolla imbricate or induplicate-
valvate, zygomorphic or actinomorphic. Endosperm present.—Ovary
1- or 2-locular. Stigmas 1 or 2, or 4. Ovules many per locule. . . 2015
— Flowers unisexual, very rarely bisexual, then stigmas 3 or 6 and
ovary 3-locular with 1 or 2 ovules per locule. Endosperm absent.—
Plants climbing or prostrate. Tendrils present. Anthers extrorse,

rarely latrorse............................... **Cucurbitaceae**

2015. Stamens 2 or 4, adnate to the corolla, free from the style. Anthers introrse or latrorse.—Flowers bisexual...................... 2016

— Stamens 2, free from the corolla, adnate to the style. Anthers extrorse.—Herbs or undershrubs. Leaves alternate. Anthers with 1 slit... **Stylidiaceae**

2016. Corolla nearly actinomorphic. Stamens 2.—Leaves opposite.... 2017

— Corolla usually distinctly zygomorphic. Stamens 4.—Disk usually present. Ovary 1-locular or incompletely 2-locular..... **Gesneriaceae**

2017. Stamens not cohering around the style. Anthers with 1 twisted theca. Disk absent. Stigma 2–4-lobed.—Flowers in cymes. Ovary incompletely to nearly completely 2-locular. N. Andes.
Columelliaceae

— Stamens cohering around the style. Thecae not twisted. Disk present. Stigma clavate to fusiform, or bifid.—Leaves dentate. Corolla valvate with a hairy ridge inside. S.E. Asia. (*Carlemanniaceae*).
Caprifoliaceae

2018. (2003). Fertile stamens more than the corolla-segments........ 2019

— Fertile stamens as many as the corolla-segments. 2022

2019. Flowers unisexual. 2020

— Flowers bisexual. 2071

2020. Leaves alternate. 2021

— Leaves opposite.—Flowers solitary or in fascicles. Style 6–10-fid. Ovules many *per ovary*. .... **Rubiaceae**

2021. Leaves usually pinnately compound. Male flowers in catkins, female flowers in a cupule. Bracts often sepaloid. Stigmas 2–4. Ovary 1-locular. Ovule 1............................... **Juglandaceae**

— Leaves simple. Flowers differently arranged. Bracts not sepaloid. Stigma 1. Ovary 2- or 3-locular. Ovules 2–4 per locule.
Symplocaceae

2022. Stamens alternipetalous................................. 2023

— Stamens epipetalous..................................... 2066

2023. Ovule 1 *per ovary*...................................... 2024

— Ovules 2–more *per ovary*................................ 2037

2024. Ovule erect... 2025

— Ovule pendulous. 2029

2025. Stigmas 3.—Tendrils or watch-spring hooks present. 2026

— Stigmas 1 or 2.—Corolla valvate. Anthers introrse. 2027

2026. Herbs. Tendrils present. Flowers unisexual. Corolla valvate. Anthers extrorse. Endosperm absent. .... **Cucurbitaceae**

— Woody plants. Watch-spring hooks present. Flowers bisexual. Corolla imbricate. Anthers introrse. Endosperm present.
Ancistrocladaceae

2027. Flowers usually in capitules. Stigma not surrounded by an involucre. 2028
— Flowers solitary, or in cymes, or in spikes, or in panicles, rarely in capitules. Stigma surrounded by a cup-shaped or 2-lobed involucre.—Corolla 5-lobed, more or less zygomorphic. Endosperm present. . . . . **Goodeniaceae**
2028. Anthers free. Style deeply divided. Endosperm present.—Leaves opposite or in whorls. Stipules present. . . . . **Rubiaceae**
— Anthers connate, rarely free, then female flowers without a distinct corolla. Style occasionally undivided, usually bifid. Endosperm absent. . . . . **Compositae**
2029. Flowers unisexual.—Leaves alternate, rarely opposite. Endosperm absent. 2030
— Flowers bisexual or polygamous.—Anthers introrse. 2031
2030. Non-resiniferous herbs. Tendrils present. Anthers extrorse. Style simple, at least at base. Embryo straight. . . . . **Cucurbitaceae**
— Resiniferous (poisonous!) trees. Tendrils absent. Anthers introrse. Styles 3, free to base. Embryo curved. . . . . **Anacardiaceae**
2031. Style 1. Stigma 1.—Flowers more or less actinomorphic. Endosperm present. 2034
— Style 1 or 3. Stigmas 3. 2032
2032. Herbs, undershrubs, or shrubs. Leaves opposite, or in whorls, or all radical. Corolla imbricate. Style 1, or 3-partite. Embryo straight.
2033
— Resiniferous (poisonous!) trees. Leaves alternate. Corolla valvate. Styles 3, free. Embryo curved.—Flowers in panicles. Endosperm absent. . . . . **Anacardiaceae**
2033. Corolla 3-lobed. Style 1. Endosperm absent.—Herbs, or undershrubs, or shrubs. Leaves opposite or all radical. . . . . . **Valerianaceae**
— Corolla 4- or 5-lobed. Style 3-partite. Endosperm fleshy.—Usually shrubs. Leaves opposite or in whorls, undivided or lobed. (*Viburnum*). . . . . **Caprifoliaceae**
2034. Leaves alternate or all radical. Epicalyx absent. 2035
— Leaves opposite or in whorls. Epicalyx present.—Herbs, rarely nonericoid undershrubs. Flowers in capitules. Corolla lobed, imbricate.
Dipsacaceae
2035. Herbs or ericoid shrubs. Flowers in capitules. Disk absent. 2036
— Non-ericoid undershrubs or trees. Flowers in axillary cymes. Disk usually conspicuous.—Corolla deeply divided, valvate. Tropics.
Alangiaceae
2036. Herbs. Corolla lobed, valvate. S. America. . . . . **Calyceraceae**
— Ericoid shrubs. Corolla deeply 5-partite, imbricate. S. Africa. (*Berzelia*). . . . . **Bruniaceae**

2037. (2023). Corolla imbricate..................................... 2038
— Corolla valvate, rarely apert............................... 2048
2038. Ovary 2 – more-locular, rarely 1-locular, then corolla actinomorphic and stipules present..................................... 2039
— Ovary 1-locular.—Leaves undivided. Stipules absent. Corolla usually zygomorphic. Style 1. Ovules many............. **Gesneriaceae**
2039. Ovary hemi-inferior.—Leaves simple. Stipules absent. Calyx 5-partite. Corolla actinomorphic, contort. Style-apex thickened, glabrous above, outer or lower side stigmatic. Ovary 2-partite or 2-locular. Ovules 2 – more per locule. 2040
— Ovary inferior, rarely hemi-inferior, then either stipules present, or corolla imbricate. Style-apex stigmatic at the summit or between the lobes....................................... 2041
2040. Stamens connate and adnate to the style-apex into a more or less capitate body. Pollen coherent into paired pollinia, each pair united by a thread-like structure ('caudicle').—Leaves above with or without a tuft of short cylindric, hair-like appendages ('colleters') at the base of the midrib. Corolla often more or less urceolate, the tube usually shorter than the lobes..................... **Asclepiadaceae**
— Stamens free, adnate to the style apex or not. Pollen free, without caudicles.—Leaves without colleters. Corolla usually rotate, or campanulate, or funnel-, or salver-shaped, the tube usually longer than the lobes. .... **Apocynaceae**
2041. Leaves alternate. 2042
— Leaves opposite or in whorls. 2046
2042. Flowers solitary, or in fascicles, or in spikes, or in racemes. Style undivided. 2043
— Flowers in capitules. Style 2-partite.—Ovules 2 per locule. S. Africa. .... **Bruniaceae**
2043. Anthers with longitudinal slits. 2044
— Anthers with apical pores.—Woody plants. Flowers solitary, or in fascicles, or in racemes. Ovules many per locule. .... **Ericaceae**
2044. Woody plants... 2045
— Herbs.—Flowers in spikes. Ovules many per locule. (also in *Campanulaceae*)..................................... **Sphenocleaceae**
2045. Flowers solitary, or in fascicles, or in spikes, or in racemes. Ovules 2 – 4 per locule...................................... **Symplocaceae**
— Flowers solitary, axillary. Ovules numerous.—Ericoid shrubs. S. Australia. (*Prionotaceae: Wittsteinia*, also in *Ericaceae*).**Epacridaceae**
2046. Stipules present. Disk present, rarely absent, then ovary 2-locular.— Leaves always undivided and entire. 2047
— Stipules absent, rarely present, then ovary 3 – 5-locular and either disk absent or leaves dentate to divided.—Woody plants, rarely her-

188

baceous. Ovary 2–6-locular. Endosperm copious. ... **Caprifoliaceae**

2047. Flowers more or less zygomorphic. Stamens unequally inserted on the corolla-tube. Ovary hemi-inferior. Ovules 2–4 per locule. Endosperm absent.—Flowers large, showy, in terminal thyrses. Fruit a loculicid capsule. Seeds not winged. N. tropical. S. America. (probably erroneously included in *Rubiaceae*). .... **Henriqueziaceae**

— Flowers usually actinomorphic. Stamens inserted at the same level. Ovary inferior. Ovules usually numerous per locule. Endosperm present. (incl. *Naucleaceae*).......................... **Rubiaceae**

2048. (2037). Flowers unisexual. Endosperm absent.—Leaves alternate.
$\qquad$ 2049

— Flowers bisexual or polygamous, rarely unisexual, then either leaves opposite or in whorls, or flowers zygomorphic and anthers introrsely or apically dehiscent. Endosperm present, rarely absent, then leaves opposite or in whorls. 2050

2049. Stipules absent. Flowers actinomorphic, rarely slightly zygomorphic, 5-, rarely 3- or 6-merous. Anthers extrorse, thecae usually tortuous.—Plants usually climbing or prostrate, usually with tendrils.
$\qquad$ **Cucurbitaceae**

— Stipules present. Flowers zygomorphic, 4-merous. Anthers latrorse.—Leaves undivided. Corolla shortly lobed. Style 3-partite, stigmas partite again. Colombia. (*Begoniella*).......... **Begoniaceae**

2050. Stigma without an involucre, but often surrounded by a ring of hairs. 2051

— Stigma with a cup-shaped or 2-lobed involucre.—Latex absent. Leaves simple. Stipules absent. Flowers 5-merous, bisexual, usually zygomorphic.................................... **Goodeniaceae**

2051. Stipules absent.—Leaves simple. Style undivided. Ovules 2–more per locule. 2052

— Stipules present, rarely absent, then either style 2-partite or ovule 1 per locule. 2062

2052. Ovary inferior, rarely hemi-inferior, then either flowers zygomorphic, or stamens free from the corolla and ovary 2–more-locular.
$\qquad$ 2053

— Ovary hemi-inferior.—Ovary 1-locular, or 2–5-locular, then either flowers actinomorphic and stamens free, or stamens adnate to the corolla. Ovules many per locule. Fruit a capsule, rarely a berry. 2059

2053. Stigma, at least after anthesis, partite, or when lobed, stem herbaceous or woody at base only, rarely undivided and more or less clavate, then *either* stem herbaceous and flowers zygomorphic, *or* flowers zygomorphic.—Latex usually present. Ovary rarely hemi-inferior, then flowers zygomorphic and stamens free from the corolla. 2054

— Stigma capitate.—Stem woody. Flowers actinomorphic. Ovary inferior. 2056

2054. Leaves usually symmetric. Inflorescences various, usually capitate, or panicles, or flowers solitary. 2055

— Leaves strongly asymmetric. Flowers in curved cincinni.—S.E. Asia to Malesia. (also in *Campanulaceae*). . . . . **Pentaphragmataceae**

2055. Flowers zygomorphic. Anthers connate. (*Lobeliaceae*).

Campanulaceae

— Flowers usually actinomorphic. Anthers free. . . . . . . **Campanulaceae**

2056. Corolla-segments either dentate to fimbriate, or tube inside with a transverse ring. New Caledonia, New Zealand. . . . . . . **Alseuosmiaceae**

— Corolla different. 2057

2057. Flowers in a terminal panicle. Stamens free from the corolla. Fruit a capsule. Réunion. (*Berenice*). **Campanulaceae**

— Flowers axillary, usually solitary. Stamens adnate to the base of the corolla. Fruit a berry. 2058

2058. Anthers with longitudinal slits. S. Australia. (*Prionotaceae*: *Wittsteinia*, also in *Ericaceae*). . . . . **Epacridaceae**

— Anthers with apical pores. Mexico to tropical S. America. (*Sphyrospermum*). . . . . **Ericaceae**

2059. Plants not twining. Latex absent. Leaves radical or alternate. Apex of stigma stigmatic. Anthers not caudate and not adnate to the stigma. 2060

— Plants usually twining. Latex present. Leaves opposite. Stigma enlarged.—Plants woody, at least at base. Flowers in cymes or in panicles, actinomorphic. Ovary 2-locular, easily separating into 2 parts. 2040

2060. Woody plants. Sepals valvate. Stamens free from the corolla. Ovary 2–5-locular.—Leaves alternate. Flowers in panicles or in umbelloid panicles. Corolla partite. Stigma capitate, 2–5-lobed. Australia, New Caledonia. 2061

— Erect herbs. Sepals imbricate. Stamens adnate to the corolla-tube. Ovary 1-locular.—Leaves radical or alternate. Flowers in cymes or in panicles. Stigma simple or 2-lobed. . . . . **Gentianaceae**

2061. Leaves glabrous beneath. Corolla lobed, with a transverse ring in the throat. Fruit a berry. (*Periomphale*). . . . . **Alseuosmiaceae**

— Leaves velvety underneath. Corolla deeply fid, throat without such a ring. Fruit a capsule. (*Escalloniaceae*: *Argophyllum*). **Saxifragaceae**

2062. (2051). Leaves alternate.—Ovule 1 per locule. Fruit a drupe or a berry. 2063

— Leaves opposite or in whorls, rarely alternate, then ovules many per locule.—Flowers usually cymose. Stamens adnate to the corolla. 2064

2063. Leaves usually compound. Stipules present, often intra-petiolar.

Flowers in umbels, or in capitules, or in spikes, or in panicles. Stamens free from the corolla. Stigmas 2–more. Ovary 5–more-, rarely 2-locular.—Petals usually calyptrate. . . . . **Araliaceae**

— Leaves undivided or lobed. Stipules absent. Flowers in cymes. Stamens adnate to the corolla. Stigma 1. Ovary 1–3-locular.—Tropics. . . . . **Alangiaceae**

2064. Plants usually woody. Ovary inferior, rarely hemi-inferior, then either style apically bifid, or ovule 1 per locule. 2065

— Herbs. Ovary hemi-inferior.—Leaves undivided. Anthers extrorse or introrse, rarely latrorse. Styles free at base, connate above. Ovary 2-locular. Ovules many per locule. (*Spigeliaceae: Mitrasacme, Mitreola*). . . . . **Loganiaceae**

2065. Leaves undivided or lobed. Anthers extrorse or latrorse. Stigma capitate or branched. Ovary usually 2-locular. (incl. *Naucleaceae*).

Rubiaceae

— Leaves deeply incised to pinnately compound. Anthers extrorse. Stigma 3–5-partite. Ovary 3–5-locular. (*Sambucaceae*).

Caprifoliaceae

2066. (2022). Corolla imbricate. Ovules many per locule. 2067

— Corolla valvate. Ovules 1–3 per locule.—Leaves undivided or absent. Fruit a drupe, or a berry, or a nut. 2070

2067. Leaves undivided. Ovary 1-locular. 2068

— Leaves digitately compound. Ovary 5-locular.—Trees. Stipules present. Flowers solitary or in fascicles. Calyx undivided or 3–5-lobed, valvate. Style undivided. Fruit a capsule, often hairy inside.

Bombacaceae

2068. Stipules absent. Calyx 5-merous. Style undivided. Flowers in racemes or panicles. 2069

— Stipules present. Flowers solitary or in cymes. Calyx 2-partite. Style 3–8-fid.—Herbs. Fruit a capsule. . . . . **Portulacaceae**

2069. Herbs or undershrubs. Calyx 5-fid. Staminodes 5. Fruit a capsule.

Primulaceae

— Woody plants. Calyx 5-lobed. Staminodes absent. Fruit a drupe or a nut. (*Maesa*). . . . . **Myrsinaceae**

2070. Plants parasitic, usually epiphytic. Leaves opposite or in whorls, sometimes absent. Ovary 1-, rarely 2-, or 3-locular. Ovule not distinct from the ovary tissue. . . . . **Loranthaceae**

— Plants autotrophic. Leaves alternate. Ovary 3-locular nearly to the apex. Ovule distinct. (incl. *Erythropalaceae*). . . . . **Olacaceae**

2071. (2019). Stamens 4–10. 2072

— Stamens 11–more. 2086

2072. Anthers 6–10. 2073

— Anthers 4.—Shrubs. Leaves opposite or in whorls, undivided.

Flowers solitary or in cymes. Style undivided. Stigma 1. Ovary 3-locular, 1 locule with 1 fertile ovule, 2 with several sterile ovules.
Caprifoliaceae

2073. Style 1, undivided. Stigma 1, undivided or lobed............. 2074
— Style 1, undivided with 2–more stigmas, or partite, or styles 2–more, free... 2081
2074. Ovary 1-locular.—Leaves alternate, rarely in whorls, simple. Stamens free from the corolla or nearly so. 2075
— Ovary 2–more-locular, sometimes apically 1-locular.—Woody plants... 2076
2075. Erect woody plants. Sepals connate. Disk present. Anthers with apical pores. Ovules few, axillary. Fruits indehiscent. .... **Ericaceae**
— Plants usually herbaceous, frequently twining. Sepals free. Disk absent. Anthers with longitudinal slits. Ovules numerous, parietal. Fruit a capsule..................................... **Loasaceae**
2076. Stamens free from the corolla, or, when adnate to it, staminodes absent; staminodes, when present, free. 2077
— Fertile stamens adnate to the middle of the corolla-tube. Staminodes connate into a tube.—Leaves alternate, undivided. Flowers 4-merous. Disk absent. Ovary 4-locular, ovules 8. Fruit dry, indehiscent. Tropical S. America. (*Lissocarpa*). .... **Ebenaceae**
2077. Leaves alternate, rarely in whorls. 2078
— Leaves opposite.—Leaves undivided. Anthers with 2 slits or pores. Ovary inferior, 5–15-locular. Endosperm absent... **Melastomataceae**
2078. Leaves digitately compound. Anthers with 1 longitudinal slit.—Calyx valvate, epicalyx often present. Ovary hemi-inferior, 5-locular. Ovules many. Fruit a capsule.............. **Bombacaceae**
— Leaves undivided. Anthers dehiscing otherwise.............. 2079
2079. Anthers dehiscing longitudinally. 2080
— Anthers with terminal pores.—Stamens usually free from the corolla, or nearly so. Disk present. Ovary 2–10-locular. Fruit indehiscent.. **Ericaceae**
2080. Calyx-segments 4 or 5, valvate or apert. Disk absent. Ovary at base 3–5-locular, apically 1-locular. Ovules many. Fruit dry, indehiscent.
Styracaceae
— Calyx 5-fid, imbricate. Disk present. Ovary completely 2–5-locular. Ovules 2–4 per locule. Fruit a drupe............... **Symplocaceae**
2081. Ovary 1-locular.—Leaves alternate, undivided. Corolla imbricate.
2082
— Ovary 2–more-locular..................................... 2084
2082. Calyx-lobes 4 or 5. Ovules not central. 2083
— Sepals 2. Ovules central.—Herbs. Stipules present. Ovary hemi-inferior. Ovules numerous. (*Portulaca*). .... **Portulacaceae**

192

2083. Plants usually herbaceous, erect or climbing, then without hooks. Ovules many, parietal.—Calyx-lobes 4 or 5. Ovary inferior. S.W. U.S., Mexico. (*Petalonyx*). .... **Loasaceae**
— Woody climbers with watch-spring hooks. Ovule 1 per locule, basal, erect.—Flowers in racemes or in panicles. Calyx and corolla 5-fid. Anthers with longitudinal slits. Style 1, undivided. Stigmas 3. Fruit dry, indehiscent with accrescent calyx-lobes. Tropical Africa to W. Malesia. .... **Ancistrocladaceae**
2084. Stem woody at least at base. Corolla valvate. 2085
— Herbs. Corolla imbricate.—Leaves radical and opposite, pinnately divided. Stipules absent. Flowers in glomerules. Anthers with 1 longitudinal slit. Styles 3–5, free. Ovary 3–5-locular. Ovule 1 per locule, pendulous. Fruit a drupe. .... **Adoxaceae**
2085. Leaves alternate, divided or compound. Stipules absent or intra-petiolar.—Ovary 2–25-locular. Ovule 1 per locule. ...... **Araliaceae**
— Leaves opposite, undivided. Stipules present, often inter-petiolar.— Ovary 4–10-locular, inferior. Ovules many per locule and flowers unisexual, or ovule 1 per locule and flowers bisexual and ovary 4-locular. (*Lasianthus*). .... **Rubiaceae**
2086. (2071). Corolla calyptrate.—Plants woody, at least at base. 2087
— Corolla connate at base only, or connate, then saucer-shaped or campanulate. 2089
2087. Fruit a drupe. Endosperm present.—Leaves alternate. 2088
— Fruit a capsule. Endosperm absent.—Leaves translucent-glandular-punctate, undivided. Style undivided. Stigma 1. Ovary inferior, 2–4-locular. Ovules many per locule.................... **Myrtaceae**
2088. Leaves undivided. Flowers solitary or in fascicles. Anthers with pores. Style undivided. Stigma 1. Ovary hemi-inferior. Ovules many per locule. Seeds long-hairy.—Stipules absent. Tropical Africa. (*Rhaptopetaleae*). .... **Scytopetalaceae**
— Leaves divided or compound. Flowers in umbels, or in capitules, or in racemes, or in panicles. Anthers with slits. Stigmas 2–25. Ovary inferior. Ovule 1 per locule. Seeds not long-hairy.—Stipules absent or intra-petiolar. .... **Araliaceae**
2089. Ovary 1-locular, rarely 3–5-locular at base.—Leaves simple or absent... 2090
— Ovary 2–more-locular. 2094
2090. Leaves well-developed. Fruit either a capsule, or dry and in-dehiscent, or a schizocarp and then sepals distinct and style 1. . 2091
— Leaves scale-like or absent, rarely well-developed, then, as usual corolla-segments and stigmas many. Fruit a berry.—Usually very succulent plants. Sepals 4–more, not clearly distinct from the

petals. Placentas 4–more, parietal. Style 1. Stigmas several.

Cactaceae

2091. Herbs. Ovary strictly 1-locular............................2092
— Woody plants. Ovary 3–5-locular at base, apically 1-locular.—
Stipules absent. Sepals 4 or 5. Style 1. Stigma 1. Disk absent. Fruit
dry, indehiscent. Placentas axillary...................**Styracaceae**

2092. Stipules usually absent. Sepals 4–7. Ovary usually inferior.....2093
— Stipules present. Sepals 2. Ovary hemi-inferior.—Style 3–8-partite.
Placenta central. Ovules many. Fruit a capsule, or dry and indehis-
cent. (*Portulaca*)................................**Portulacaceae**

2093. Corolla-segments 4 or 5. Stigmas 1 or 4. Placentas several, parietal.

Loasaceae

— Corolla-segments many. Stigmas 4–12. Placenta central.—Plants
more or less fleshy. (*Mesembryanthemum*)..............**Aizoaceae**

2094. Anthers with pores.—Woody plants. Stamens twice as many as the
corolla-segments, free from these. Style 1. Stigma 1..........2095
— Anthers with longitudinal slits............................2096

2095. Leaves opposite. Corolla-segments imbricate......**Melastomataceae**
— Leaves alternate. Corolla-segments valvate.—Flowers in corymbs.

Ericaceae

2096. Style 1, undivided. Stigma 1, or capitate and/or 3–8-lobed.—Woody
plants. Leaves alternate, undivided. Corolla imbricate or plicate.
Fruit indehiscent, rarely a capsule........................2097
— Style 1, partite or divided, stigmas several, or styles many.....2100

2097. Calyx valvate or apert. Endosperm absent (unrecorded for
Napoleonaeaceae)......................................2098
— Calyx imbricate. Endosperm copious.—Ovules 2–4 per locule. Fruit
a drupe.................................**Symplocaceae**

2098. Corolla plicate, 20–40-rayed, margin dentate.—Fruit a berry or a
non-operculate capsule....................................2099
— Corolla imbricate, segments 4–6, connate at base only.—Stamens
many. Fruit a berry or a woody operculate capsule...**Lecythidaceae**

2099. Flower perigynous. Sepals connate, apert, many. Stamens many.
Style filiform, stigma simple. Fruit a capsule. Brazil. (*Asteranthaceae*)................................**Lecythidaceae**
— Flower epigynous. Sepals free, 5, valvate. Stamens 10–20, staminodes many. Fruit a berry. W. Africa. (*Napoleonaeaceae*).

Lecythidaceae

2100. Leaves simple, when divided leaves submerged. Flowers solitary.
Ovules 2–many per locule.—Corolla imbricate..............2101
— Leaves partite to compound. Flowers in umbels, or in capitules.
Ovule 1 per locule.—Terrestrials. Corolla often valvate. Stigmas
several...**Araliaceae**

194

2101. Herbs, usually aquatic. Leaves radical. Corolla-segments and styles many. 2102
— Woody, terrestrial plants. Leaves alternate. Corolla-segments 5. Style either 1, 3-partite (*Ternstroemiaceae*: *Anneslea*, E. Asia), or styles 3. (*Visnea*, Canary Isl.).—Endosperm scanty to absent.

Theaceae

2102. Sepals hypogynous. Corolla epigynous. Stamens adnate to the corolla. S.E. Asia. (*Barclayaceae*). . . . . **Nymphaeaceae**
— Sepals and petals epigynous. Stamens free from the corolla, at least the outer ones. Tropical E. Asia and S. America. (*Euryalaceae*).

Nymphaeaceae

PARASITES AND SAPROPHYTES

2103. (158). Plants herbaceous, terrestrial or twining. Stems with scales, distinct leaves absent. 2104
— Plants woody, or thick-fleshy, or herbaceous, in the latter case either green leaves present, or plants epiphytic, hemi-parasitic and erect. 2108
2104. Stems twining. Parasites with haustorial organs on the stems. . . 2105
— Stem erect. Saprophytes. 2106
2105. Petals connate. Filaments adnate to the corolla-tube, alternipetalous. Anthers longitudinally dehiscing with slits. Fruit a capsule. Seeds 1–4. (*Cuscutaceae*). . . . . **Convolvulaceae**
— Petals free. Filaments free from the corolla, in 3 whorls of three and epipetalous, or more. Anthers dehiscing with an apical valve. Fruit 1-seeded, surrounded by a fleshy receptacle.—Sepals 3. (*Cassytha*).

Lauraceae

2106. Scales on the stem opposite.—Corolla-lobes contort or 4-lobed (2 inner and 2 outer lobes). Filaments adnate to the corolla-tube.

Gentianaceae

— Scales on the stem alternate.—Petals or corolla-lobes imbricate, not contort. Filaments free from the corolla. 2107
2107. Sepals 2–6. Petals absent or 3–6, free or connate. Ovary 1–6-locular. Ovules many, axillary or parietal. . . . . **Monotropaceae**
— Sepals 5. Petals 3. Ovary 2-locular. Ovule 1 per locule, pendulous.—Indomalesia, Australia. (*Salomonia*). . . . . **Polygalaceae**
2108. Plants with chlorophyll, rarely without, then flowers either on branched stems, or in compound inflorescences. Usually epiphytic hemi-parasites. Fruit 1-seeded. 2109
— Plants without chlorophyll, usually parasiting on roots, if epiphytic, then flowers emerging solitary from the host's branches. Fruit many-

seeded. 2113

2109. Flowers unisexual, the male flowers consisting of a group of up to 3 stamens. Fruit dry, with 3 feather-like bristles.—Epiphytic, shrubby, green parasites on *Nothofagus*. Temperate S. America.

Myzodendraceae

— Flowers bisexual or unisexual, in the latter case the male flowers either with a perianth, or (*Antidaphne*) consisting of a group of 4 stamens. Fruit usually fleshy, without feather-like bristles. 2110

2110. At least the bisexual or female flowers with a rim-like calyx (calyculus) below the corolla.—Flowers usually brightly coloured and usually bisexual, if flowers unisexual then plants dioecious.

Loranthaceae

— Calyx or calyculus absent.—Plants monoecious or dioecious. Flowers usually inconspicuous, greenish. 2111

2111. Leaves usually decussate. Flowers in cymes or produced from the stem, not the leaf-axils (Tropical America, West Indies: *Dendrophthora, Phoradendron*). Anthers usually sessile or cohering. **Viscaceae**

— Leaves usually alternate. Flowers in axillary or terminal racemose inflorescences. Anthers neither sessile, nor cohering. 2112

2112. Plants attached by means of large, distinct primary haustoria, sometimes also with secondary haustoria on creeping roots. Fruitwall without conspicuous longitudinal fibres. S. America, Mexico, Caribbean. . . . . **Eremolepidaceae**

— Plants without a distinct primary haustorium. Branches either leafy or with scales and then originating from endophytic parts. Fruitwall with conspicuous longitudinal fibres. S.E. Asia, New Guinea.

Santalaceae

2113. Flowers distinctly zygomorphic. 2114

— Flowers actinomorphic. 2115

2114. Ovary 2-locular.—Primary haustorium present or absent. Subterranean stem often branched. Old World. . . . . **Scrophulariaceae**

— Ovary 1-locular, rarely incompletely divided into locules.—Primary haustorium present. Subterranean stem usually simple. World wide.

Orobanchaceae

2115. Ovule 1, or indistinct and fused with the ovary wall.—Flowers unisexual, in club-shaped or disk-shaped inflorescence. **Balanophoraceae**

— Ovules more than 10 and distinct.—Flowers unisexual or bisexual.

2116

2116. Flowers either in inflorescences, or solitary and emerging apparently directly from the host, then rhizome-like subterranean parts absent.

2117

— Flowers solitary, emerging from a coarse, rhizome-like, subterranean part of the parasite.—Madagascar, S. Africa, S. America.

Hydnoraceae

2117. Flowers in branched inflorescences. Stamens adnate to the corolla, with distinct filaments. Ovules less than 16.—Dry habitats in America...**Lennoaceae**

— Flowers either solitary or in simple spikes. Anthers sessile on a central column, without distinct filaments. Ovules numerous.

Rafflesiaceae

GLOSSARY

Abaxial Facing away from the axis.

Achene A one-seeded, dry, indehiscent fruit with the seed free from the pericarp.

Actinomorphic Regular: a flower with radially arranged (sub-)equal perianth-segments.

Adaxial Facing the axis.

Adnate *Of organs*: fusion of non-homologous ones (petals with stamens, etc., see *connate*); *of anthers*: more or less fused with the filament and not movable freely and independently from the latter (see *versatile*).

Aestivation The way in which the floral parts are placed in bud.

Alternate *Of leaves*: attached solitary and spaced along the axis.

Alterni- a prefix: alternating with, as in *alternipetalous stamens*: stamens alternating with the petals.

Anatropous Ovules with the raphe so adnate to the straight nucellus that the micropyle is next to the funicle.—**Plate 3**.

Androgynophore A stalk supporting both the stamens and the pistil(s).

Androphore A stalk supporting the stamens.

Annual *Of herbs*: completing the full cycle of germination to fruiting within the year and then dying.

Anther The part of the stamen containing the pollen, usually bilocular and the locules ('*thecae*') connected by the connective.

Antidromous *Of stipules*: connate on one side, but not over the petiole (then *intra-petiolar*, q.v.), leaving a ring-like scar around the twig, as in *Ficus, Platanus*.

Apert Margins of the perianth-segments not touching each other in bud, except perhaps at the very base.

Apocarpous Composed of 2 or more mutually free carpels.

Apotropous An anatropous ovule with the funicle facing away from the placenta when pendulous, to next to it when erect. (cf. *epitropous*).—**Plate 2: 2, 4**.

Aril A usually fleshy or membranous cover of the seed originating from the hilum, or funicle, or placenta, or micropylar area.

Articulated Provided with a joint or pre-formed breakage-point (in pedicels, petioles, or fruits).

Ascending *In stems*: prostrate at base, becoming erect upwards; *of ovules*: with the funicle pointing upwards.—**Plate 2**: **3**, **4**.

Asymmetric Not divisable by any plane into two (sub-)equal parts.

Atropous *Of ovules*: funicle, nucellus, and micropyle in one line; a straight (*orthotropous*) ovule.—**Plate 3**.

Auricle A lateral (usually rounded) appendage (in a leaf at the base of the blade or petiole itself, not to be confused with the stipules, q.v.).

Autotrophous A green, non-parasitic, non-saprophytic plant.

Awn A strong bristle or bristle-like structure.

Axillary Standing in an axil; *of ovules*: attached along the central axis in a loculed ovary.

Basifix *Of anthers*: filament attached at or near the base of the anther.

Berry A fleshy or juicy fruit, indehiscent, endocarp not indurated, seeds not in distinct locules.

Bi- A prefix: two, as in *bilabiate*: with two lips.

Biennial *Of herbs*: completing the full cycle of germination to fruiting in more than one, but not more than two years and then dying.

Bisexual Having both fertile stamens and pistils in one flower.

Bract Any modified, usually reduced leaf, usually the ones subtending a flower or (part of) an inflorescence.

Bracteole One or more bracts on a pedicel. (*Note*: to be present on the pedicels of *all* flowers, otherwise to be regarded as bracts).

Bulb A short, usually subterranean part of the plant composed of thickened scales.

Calycoid Resembling a calyx.

Calyptra Cap-shaped, see *closed*.

Calyx The outermost floral envelope (but cf. *epicalyx*), usually smaller and drier than the next inner one (*corolla*), and more or less green.

Campanulate Bell-shaped: tube about as long as wide, gradually enlarged into the limb.

Campylotropous A form of *anatropous*, q.v.—**Plate 3**.

Capitate Head-shaped, as the knob of a pin; *of flowers*: in capitules.

Capitule An inflorescence with more or less sessile flowers on a common receptacle, surrounded by an involucre (if not, see *glomerule*).

Capsule A dry fruit, dehiscing in various ways, derived from 2 or more carpels.

Carpel A leaf-derived organ bearing ovules. (An ovary is considered to be composed of 1–more carpels).

Caruncle A wart or protuberance on the seed, see also *obturator*.

Caryops A one-seeded, dry, indehiscent fruit with the pericarp adnate to the testa.

Catkin   A dense raceme or spike, usually pendulous, with minute unisexual flowers, falling as a whole.

Cf.   Compare, see.

Chalaza   *Of ovules or seeds*: the place where the nucellus meets the integuments; opposite the cotyledons.—**Plate 2: 5**.

Cincinnus   A cymose, dichotomous inflorescence resembling a raceme, in which the apparent main axis is in fact composed of secondary ones, i.e. an actually lateral branch forms the internode. Note the presence of a bract or leaf opposite to the flower and not subtending it, as in truly racemose inflorescences.

Clavate   Club-shaped.

Closed   *In aestivation*: all parts connate, either separating at anthesis, or deciduous together because of a transverse suture as a calyptra.

Coherent, cohering   *Of organs*: glued, but not fused together, and to be separated with caution without tearing.

Collateral   Placed side by side, as in ovules.

Columella   *In fruits*: the persistent central axis after dehiscence.

Compound   Consisting of free parts: leaflets in leaves, partial inflorescences in inflorescences, etc.

Cone   A spike-like inflorescence with large, indurating bracts; the ultimate pseudocarp; a flower, inflorescence, or fruit resembling this.

Connate   *Of organs*: fusion of homologous ones, e.g. petals among themselves, etc.; see *adnate*.

Connective   The tissue between the locules ('*thecae*') of the anther (usually very inconspicuous).

Contort   Margins of the perianth-segments overlapping each other so that one part is inside, the other outside, and none is completely inner- or outermost. (Note: this state, unless expressly stated is usually included in *imbricate*, q.v.).

Cordate   At base with an acute incision between two rounded lobes, generally also with a more or less acute apex.

Corniculate   With horn-shaped appendages.

Corolla   The inner-most floral envelope (but cf. *corona*), usually larger, more flaccid than the outermost one (*calyx*), and usually coloured (not green).

Corolloid   Resembling a corolla.

Corona   One, rarely two whorls of petaloid, or thread- or horn-like, etc. appendages between the corolla and the stamens, of corolloid or staminodial origin, as in *Narcissus*, *Passiflora* (not to be confused with the lobes of a *disk*).

Corymb   An inflorescence, usually a raceme, in which the flowers through unequal pedicels are in one (horizontal) plane.

Cotyledon The first leaf or leaves of the embryo, usually present in the seed.

Crenate Of a margin with small, sharp incisions and rounded intermediate teeth.

Cupule Connate, indurated bracts subtending or enveloping a flower or an inflorescence, as in *Fagaceae*.

Cyme A cymose inflorescence, especially one with equally developed lateral branches.

Cymose *Of an inflorescence*: branched with flowers terminating each axis; determinate.

Decussate In pairs that alternate at right angles, organs thus in four rows.

Dehiscent or **dehiscing** Opening at maturity to release the contents (pollen, seeds).

Dentate *Of a margin*: with small, blunt incisions and sharp teeth.

Descending *Of ovules*: with the funicle pointing downwards.—**Plate 2: 1, 2**.

Dichasial *Of an inflorescence*: cymose with opposite branches.

Dichotomous Divided into two equal parts.

Didynamous *Of stamens*: consisting of two unequally long pairs.

Dioecious Male and female flowers on different plants.

Disk A more or less pronounced outgrowth of the receptacle without vascular traces, ring-, cushion-, cup-shaped, etc., sometimes divided into lobes or separate bodies, or a unilateral one; generally with a nectar-secreting function.

Divaricate Divergent with an obtuse angle, usually approaching 180°.

Dorsal *Generally*: abaxial; *of a raphe*: on the side of the ovule facing away from the placenta.—**Plate 2: 2, 3**.

Dorsifix *Of anthers*: attached about halfway the length to the filament.

Drupe An indehiscent fruit with a membranous to leathery exocarp, a more or less fleshy mesocarp and a strongly indurated, woody to stony endocarp.

E.g. For example.

Ellipsoid Elliptic, but tri-dimensional.

Elliptic A two-dimensional shape, in which the length is between one and two times the width with the greatest width about the middle.

Emarginate Notched.

Embryo The rudimentary plant present in a mature seed.

Endo- A prefix: the inner . . ., as in *endocarp*, the inner layer of the pericarp, and in *endotesta*, the inner layer of the testa.

Endosperm The nutritive tissue within the seed (not of the embryo proper), usually surrounding the embryo or to one side of it (here inclusive of *perisperm*).

Entire An even margin; without any incisions or teeth.

Epi- A prefix: 1) before, as in *epipetalous stamens*: stamens inserted before the petals (not necessarily adnate to them!); 2) upon, as in *epiphyte*; 3) on, or above, as in *epigynous*; 4) next to, as in *epitropous*.

Epicalyx An involucre of a single flower resembling an outer calyx next to the actual one.

Epigynous Sepals, petals or tepals and stamens inserted on or above the plane through the apex of the ovary (which may be superior to inferior).—**Plate 1: 5, 6**.

Epimatium The ovule-bearing scale in *Coniferales*.

Epiphyte A plant growing upon an other and not rooting in the soil, usually non-parasitic.

Epitropous An anatropous ovule with the funicle next to the placenta, when pendulous, or facing away from it, when ascending (cf. *apotropous*).—**Plate 2: 1, 3**.

Equitant *Of leaves*: distichous and with overlapping leaf-bases, as in *Iris*, *Zingiber*.

Exduplicative *In aestivation*: valvate with the margins folded outwards.

Exo- A prefix: the outer . . . , as in *exocarp*, the outer layer of the pericarp, and in *exotesta*, the outer layer of the testa.

Extra- A prefix: outside, as in *extra-staminal*: outside the stamens.

Extrorse *Of anthers*: dehiscing abaxially (check in bud!).

Fascicle A group of leaves or pedicelled flowers (cf. *glomerule*), apparently originating from the same point or area of a branch (cf. *umbel*).

Fertile Provided with functional sexual parts (pollen or ovules well-developed and capable of producing seeds).

-fid A suffix: divided to about half-way the midrib.

Filament The stalk of the anther.

Follicle A dry fruit, derived from a single carpel and dehiscing along one suture.

Funicle The stalk of the ovule.—**Plate 2: 5**.

Fusiform A tri-dimensional shape, terete and tapering at both ends.

Globose Ball-shaped.

Glomerule A cluster of sessile, usually minute flowers, not surrounded by an involucre (cf. *capitule*).

Glume A more or less scarious bract subtending a specialized inflorescence, as in the spikelet of a grass.

202

Gynobasic *Of styles*: attached near or to the base of the ovary.

Gynophore A stalk supporting the pistil(s).

Hastate A shape with at base two divergent, acute lobes.

Haustorium A sucker of parasitic plants.

Hemi- A prefix: partly, as in *ovary hemi-inferior*: ovary partly adnate to the hypanthium and partly free from it.—**Plate 1**: **3**.

Hemitropous An anatropous ovule with a medially attached funicle and a terminal micropyle at a right angle to the latter.—**Plate 3**.

Herb Plant, non-woody, or woody at base only, above-ground stems usually ephimerical.

Hilum The place where the ovule or seed is or was attached to the funicle or placenta.

Hispid Provided with stiff, rigid hairs or bristles.

Hypanthium An enlarged receptacle with a more or less well-developed part between the ovary and the insertion of the perianth-segments; from the outside of the flower the difference between the hypanthium and the calyx is often obscure.

Hypogynous The sepals, petals or tepals and usually also the stamens inserted below or at the plane of insertion of the ovary. (*Note*: there may be a more or less developed receptacle with or without a disk; the ovary is always superior; the stamens may be inserted on the petals, whereby the flowers appear to be epi- or perigynous.—**Plate 1**: **1**, **2**.

Imbricate Overlapping each other by their margins, especially used for the aestivation. (*Note*: unless stated incl. *contort*, then specifically: one or two parts outermost, one or two innermost, the other(s) partly covered, partly covering).

Imparipinnate Pinnately compound with an odd number of leaflets, usually with a terminal one.

Indument The hair-like covering of an object.

Induplicative *In aestivation*: valvate with the margins folded inwards.

Inferior *Of the ovary*: completely fused with the hypanthium, at most with a free summit, if less adnate, see *hemi-*.—**Plate 1**: **5**.

Integument *Of an ovule*: its envelope(s).—**Plate 2**: **5**.

Inter- A prefix: between, as in *inter-petiolary*: between the petioles.

Intra- A prefix: within, as in *intra-petiolary*: within the axil, but abaxial to the axillary bud or branch; *intra-staminal*: within the whorl of the stamens.

Introrse *Of anthers*: dehiscing adaxially (check in bud!).

Involucre A usually bract-like structure surrounding a flower or an inflorescence (as in *Compositae*), or another organ (as the stigma in *Goodeniaceae*).

203

Irregular *Of a flower*: not to be divided into any (sub-)equal parts; asymmetric. (Usually only the perianth-segments are considered of importance).

Lanceolate A two-dimensional shape, in which the length is between three and five times the width with the greatest width about the middle.

Latex A milky juice exudated when cut, as in *Euphorbia*, *Hevea*.

Latrorse *Of anthers*: dehiscing laterally (check in bud!).

Lepidote Covered by a more or less stellate, scurfy indument.

Liana A usually woody climber without specialized climbing-organs (as in *vines*).

Ligulate Tongue-shaped; provided with a ligule.

Ligule A variously shaped appendage internal to the base of leaf-blades, or petioles, or perianth-segments.

Limb The free parts of a connate calyx or corolla, distinct from the tube.

Linear A two-dimensional shape, in which the length is more than ten times the width with the greatest width about the middle.

Linear-lanceolate A two-dimensional shape, in which the length is between five and ten times the width with the greatest width about the middle.

Lip One or more exceptionally well-developed perianth-segments, in clear contrast to the other ones of the same envelope, as in most orchids.

Lobed Divided to less than half-way the midrib (e.g. of *leaves*), or shallowly incised (e.g. of *stigmas*).

-locular A suffix: the number of locules. (*Note*: minute and obviously reduced ones devoid of ovules or seeds are not to be counted).

Locule A more or less closed cavity, containing the pollen in anthers and the ovules in ovaries. An *incomplete* locule of an ovary is one, where the septs are not completely developed and/or fused (*incomplete septs*) and one may pass from one locule to another. Locules which are incomplete at their very top have been considered as complete by Thonner.

Loculicide *Of capsules*: dehiscing between the septs or placentas into the locule.

Lomentaceous A fruit: at maturity transversely dehiscent into parts (cf. *schizocarp*).

Mericarp Part of a schizocarp.

-merous A suffix: divisable by the same basic number, e.g. 5-merous: sepals 5, petals 10, stamens 15 (the number of carpels and their style(s) or stigma(s) is usually of no importance).

Mesocarp *Of fruits*: the middle layer of the pericarp.

Micropyle The opening between the integuments of an ovule. A microscope is usually needed to observe this and/or some dose of fantasy. In the seed the radicle apparently always points towards the micropyle!—**Plate 2**: **5**.

Monoecious Male and female flowers on the same plant.

Mucro A sharp, usually suddenly constricted terminal point.

Mucronate Having a mucro.

Naked Devoid of an envelope.

Nigrescent Becoming black or dark in drying.

Nucellus The kernel of an ovule, usually surrounded by integuments, from which the embryo (and the endosperm) is formed.—**Plate 2**: **5**.

Nut A dry indehiscent fruit with a more or less indurated pericarp and a single seed.

Ob- A prefix: the other way around, as in *obovate*: ovate but widest *above* the middle.

Oblong A two-dimensional shape, in which the length is between two and three times the width with the greatest width about the middle.

Obturator A wart-like protuberance of the placenta, covering the micropyle, as in many *Euphorbiaceae*.

Orthotropous See *atropous*.

Ovary The lower part of the pistil containing the ovule(s).

Ovate A two-dimensional shape, in which the length is between one and two times the width, with the greatest width below the middle.

Ovoid Ovate, but tri-dimensional.

Palea A usually scarious bract of a common receptacle (as in *Compositae*) or the adaxial involucral bract in the spikelets of *Gramineae*.

Palmate With parts or ramifications in one plane which originate more or less from one place. (Usually incl. *pedate*).

Palmati- A prefix: palmately so.

Panicle A compound inflorescence with a main axis and at least secondary branches (usually incl. *thyrse*, specifically: main and lateral axes branched in the same way, either racemose, or cymose).

Papilionaceous *Of flowers*: zygomorphic and imbricate with one wide, upper segment, two narrower lateral ones and two narrower lower ones, the latter usually coherent or connate by their margins; as in the *Papilionaceae*.

Parasite A plant growing and feeding upon another, usually lacking chlorophyll. A *hemi-parasite* is partly parasitic, partly autotrophous, and has chlorophyll.

Parietal *Of ovules*: attached to the outer wall of the ovary; placenta sometimes excurrent or ridge-shaped.

Paripinnate Pinnately compound with an even number of leaflets, usually without a terminal one.

-partite A suffix: divided to more than half-way the midrib, but not yet compound.

Pedate With parts or ramifications in one plane, where the larger ones originate from the basal side-nerves, the next larger from the basal side-nerves of these, and so on, superficially resembling *palmate* and usually included there.

Pedati- A prefix: pedately so.

Pedicel The flower-stalk without bracts, sometimes with bracteoles.

Peduncle The stalk of the inflorescence: the axis between the last true leaf and the first branch (and bract) of the inflorescence.

Peltate Round and with a stalk or attachment somewhere on its surface, usually about the middle.

Perennial *Of herbs*: not dying after flowering and fruiting (here used incl. *biennial*).

Perianth The floral envelopes, calyx and corolla, or the floral envelope, when these cannot be distinguished.

Pericarp The fruit-wall.

Perigynous Sepals, petals or tepals and usually also the stamens inserted between the plane of insertion of the ovary and its apex, i.e. more or less around the ovary on a more or less well-developed hypanthium. (The ovary may be superior to hemi-inferior).—**Plate 1: 3, 4.**

Perisperm See *endosperm*.

Petal Free segment of the corolla.

Petaloid Resembling a petal.

Petiole The leaf-stalk.

Petiolule The stalk of a leaflet.

Phylloclade A widened, flattened and green axis, resembling a leaf.

Pinnate With parts or ramifications in one plane, which are placed along a central axis, as in a feather.

Pinnati- A prefix: pinnately so.

Pistil The female organ of a flower, composed of one or more carpels.

Pistillode A reduced pistil, without developed ovules.

Pitcher A flask-shaped to tubular modified leaf, as in *Nepenthes, Sarracenia*.

Placenta The part of the carpel which bears the ovule(s).—**Plate 2: 5.**

Plicate Folded lengthwise with pleats.

-plinerved A suffix: number of (sub-)equal nerves, as in *triplinerved*: with three (sub-)equal main nerves originating from the base of the blade.

Pod A dry fruit derived from a single carpel, dehiscing along the dorsal and ventral sutures; seeds attached dorsally.

Pollinium A body composed of all the pollen of an anther-locule, as in *Asclepiadaceae, Orchidaceae*.

Polygamous Some flowers unisexual, others bisexual on the same or different plants.

Pseudo- A prefix: resembling, as in *pseudocarp*: apparently a fruit, but composed of carpels and other parts of the flower or inflorescence, as in *Ficus, Fragaria*.

Raceme An inflorescence with a simple, elongated rachis and pedicelled flowers. (A raceme is not necessarily racemose!).

Racemose Of an inflorescence: branched without terminal flowers; indeterminate. (A racemose inflorescence is not necessarily a raceme!).

Rachis The main axis of a compound leaf or inflorescence.

Radiating Patent to all sides; *in inflorescences*: the outer flowers with a larger perianth than the inner.

Radicle The first root of the embryo, usually present in the seed. N.B.: The radicle apparently always points towards the micropyle!

Raphe *In ovules and seeds*: the vascular bundle between the nucellus and the funicle; the general area around it.—**Plate 2: 5.**

Receptacle The shortened axis of the flower, often punctiform or disk-like (cf. *hypanthium*); the *common receptacle* is the shortened axis of an inflorescence (as in *Compositae*).

Resinous Containing resin, a kind of latex usually becoming sticky or solid after contact with air, as in *Anacardiaceae* (poisonous!), *Pinus*.

Reticulate Net-shaped, e.g. of venation: veins in an irregular network shaped by the numerous interconnecting branches.

Rhizome Rootstock, part of the stem resembling a root, not covered by scales, more or less elongated and horizontal, producing shoots at one end.

Rotate *Of the corolla*: the parts spreading out in one plane from the axis; wheel-shaped.

Ruminate *Of endosperm*: intrusion of the testa into the endosperm, which then in transection resembles the pattern of a cow's tooth, as in a nutmeg.

Sagittate A shape with at base two retrorse, acute lobes.

Salver-shaped A shape: with a narrow tubular tube and a small, spreading limb.

Saprophyte A plant without chlorophyll living exclusively upon dead organic material (actually through a fungus in its basal tissues). Many plants are hemi-saprophytic, but then have chlorophyll.

207

Scale   Any thin scarious organ, either a reduced leaf, or a much flattened hair.

Scape   A peduncle, usually originating from the base of the plant, without leaves, at most with some bracts.

Scarious   Thin, dry, translucent and pale.

Schizocarp   A usually dry fruit, which splits up longitudinally into non- or tardily dehiscent parts (*mericarps*). (cf. *lomentaceous*).

Sclerenchyma   Tissue composed of thick-walled cells.

-sect   A suffix: divided to about the midrib.

Secund   *Of branches*: oriented to one side, often curving down.

Segment   Part of a structure, e.g. the lobe of a connate corolla, but also a free petal.

Sepal   Free segment of the calyx.

Sepaloid   Resembling a sepal.

Sept   The partition dividing an ovary or fruit into locules. *True septs* originate from the margins of carpels, *false septs* do not. (cf. *locule*).

Septicide   *Of capsules*: dehiscing through the septs or placentas.

Septifragous   *Of capsules*: when the valves break away from the persistently connate septs or placentas.

Serial   Placed on above the other, as in *ovules*.

Serrate   *Of a margin*: with small, sharp incisions and teeth.

Sessile   Without a stalk; in *anthers*: without filaments; in *stigmas*: without styles.

Sheath   *Of leaves*: the broadened base of a blade or petiole, usually enveloping the internode for some length.

Shrub   Woody plant without a distinct main stem, therefore usually not very high and much-branched.

Silique   A bi-locular fruit composed of two carpels, usually dehiscing with two valves, as in *Cruciferae*.

Simple   *Of a leaf*: entire to divided, but not compound; of a perianth: parts (sub-)equal, not differentiated into calyx and corolla.

Spadix   A spike-like inflorescence with an unbranched, usually thick rachis and more or less minute flowers imbedded in it, the whole generally subtended by a spathe.

Spathaceous   A structure resembling a spathe.

Spathe   An enlarged bract enclosing a (partial) inflorescence or single flower.

Spathulate   A two-dimensional shape with a broadened part (blade) and a stalk-like one (claw), as in a ping-pong-bat.

Spike   An inflorescence of a single rachis with more or less sessile flowers.

Spikelet   A small specialized spike (as in *Gramineae*).

Spine   An indurated, sharp object not derived from an organ, and therefore usually irregularily distributed (cf. *thorn*).

Spur A tubular appendage of one or more perianth-segments (usually the corolla).

Stamen The male organ of a flower.

Staminode A reduced stamen without pollen.

-stichous A suffix: in rows or ranks (usually of leaves).

Stigma The usually papillose or glandular part of the style for the receival of the pollen.

Stigmatic Having or resembling a stigma.

Stipel Stipule-like appendage at the base of a leaflet (in unifoliolate leaves inserted on the petiole, not on the stem!).

Stipitate Having a stalk or stipe, usually of an ovary or fruit.

Stipule A paired leaf-like, scale-like, spiny, glandular, bristle-shaped, etc. structure on both sides of the leaf-base or petiole, inserted on the axis; sometimes very early fugacious and then leaving a more or less distinct scar (check young shoots!).

Style The usually narrowed part of the pistil between the ovary and the stigma.

Sub- A prefix: more or less, nearly, as in *sub-equal*.

Subulate Awl-shaped: narrow, terete, and acute.

Succulent Juicy, fleshy, as the stem of *Cactaceae*.

Superior *Of the ovary*: inserted only by its base on the receptacle, but otherwise free from it.—**Plate 1: 1, 2, 4, 6**.

Symmetric Divisable by one or more planes into two or more (sub-)equal parts.

Syncarp A compound fruit originating from several, originally free carpels, as in *Magnolia, Morus*.

Syncarpous Ovary composed of several connate carpels. (A syncarpous ovary does **not** produce a syncarp!).

Tendril A long, slender, usually watch-spring-like, coiled organ derived from an axis, or leaf, or parts of these.

Tepal Free segment of a perianth not differentiated into a calyx and a corolla.

Terete Cylindric and elongated.

Ternate In threes.

Testa The more or less indurated skin of the seed enclosing the endosperm and embryo; the seed-coat.

Theca The locule of an anther.

Thorn An indurated, sharp object derived from an organ, e.g. a branch, a stipule, a leaf, and therefore more or less regularily distributed. (Cf. *spine*).

Throat The general area between tube and limb.

Thyrse A compound inflorescence with mixed types of branching: the main ones racemose, at least the ultimate ones cymose.

Tree Woody plant with a single distinct stem, generally fairly high.

Tri- A prefix: three, as in *tri-foliolate*: with three leaflets.

Tube The fused, usually elongated part of connate sepals, petals, tepals, or filaments.

Tuber A short, thickened part of the root or stem without scales.

Umbel An inflorescence in which the pedicels or secondary axes originate from one point on the top of the peduncle.

Unarmed Without spines or thorns.

Undershrub A small shrub, often partially herbaceous, the ends of the branches often dying during winter or dry season.

Unguiculate Claw-like, or having such appendages; cf. *spathulate*.

Unifoliolate A compound leaf reduced to a single leaflet, usually recognizable by the articulated 'petiole', actually a petiolule and a petiole.

Unisexual *Of flowers*: with one sex only, either the anthers with pollen, or the ovary with ovules. (Pistillodes or staminodes may be present!).

Urceolate A shape: inflated and contracted at the mouth like an urn or pitcher.

Utricle An irregularily or non-dehiscent fruit or seed enclosed in a loose, membranous pericarp or bract.

Valvate Touching each other with the margins but not overlapping; dehiscing by valves. In aestivation usually inclusive of *induplicative* (q.v.).

Valve A lid or segments of an anther or capsule after dehiscence.

Ventral Adaxial; *of a raphe*: on the side of the ovule facing to the placenta.—**Plate 2: 1, 4**.

Versatile *Of anthers*: attached with a usually small joint to the filament and freely and independently movable. (Cf. *adnate*).

Verticillate In a whorl.

Vine A usually woody climber with specialized climbing-organs, e.g. tendrils, hooks, adventitious roots, etc.

Virgate A broom-like habit, more or less densely branched with stiff, ± erect branches, leaves usually small.

Viviparous Seed germinating while still attached to the plant, as in *Rhizophoraceae*. (*Proliferous*: reproducing vegetatively with the plantlets, not derived from the seed, developing on the mother-plant before falling off).

Zygomorphic A flower which can be divided into two (sub-)equal parts by one plane only, as in an orchid; bilateral symmetric. (Usually only the perianth-segments are considered of importance).

FLOWER: POSITION OVARY VERSUS RECEPTACLE

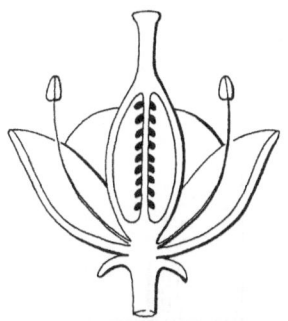

1. FLOWER HYPOGYNOUS
   OVARY SUPERIOR

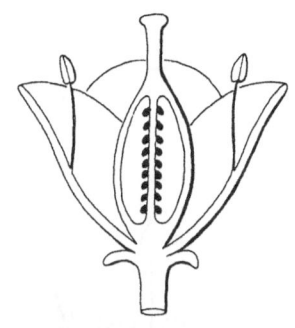

2. FLOWER HYPOGYNOUS
   STAMENS INSERTED ON THE
   COROLLA
   OVARY SUPERIOR

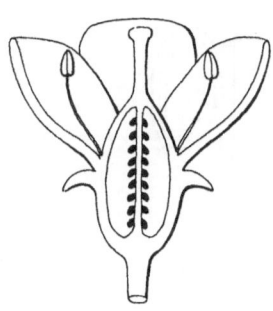

3. FLOWER PERIGYNOUS
   OVARY HEMI-INFERIOR

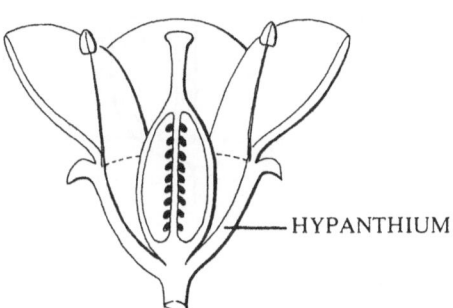

HYPANTHIUM

4. FLOWER PERIGYNOUS
   OVARY SUPERIOR

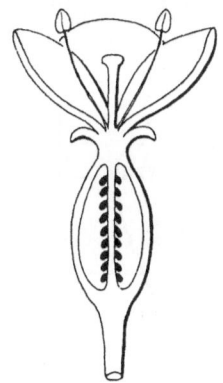

5. FLOWER EPIGYNOUS
   OVARY INFERIOR

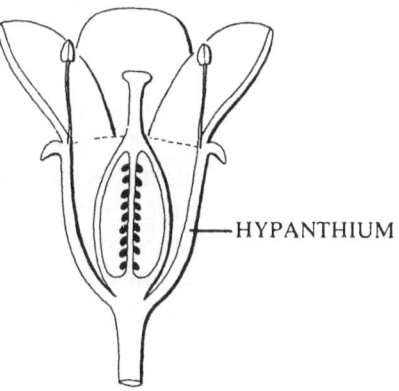

HYPANTHIUM

6. FLOWER EPIGYNOUS
   OVARY SUPERIOR

Plate 1

211

OVULES: POSITION VERSUS PLACENTA

	EPITROPOUS	APOTROPOUS
DESCENDING (pendulous)	 1	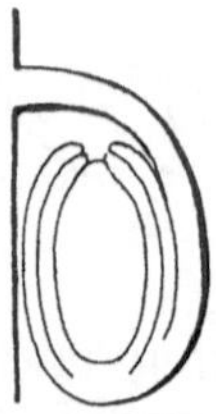 2
ASCENDING	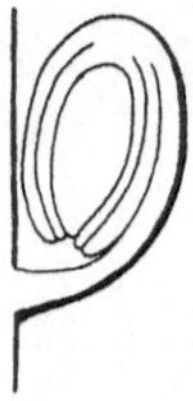 3	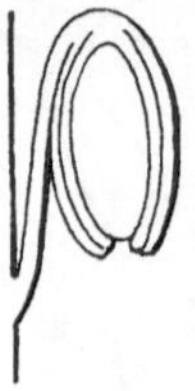 4

1 & 4: RAPHE VENTRAL

2 & 3: RAPHE DORSAL

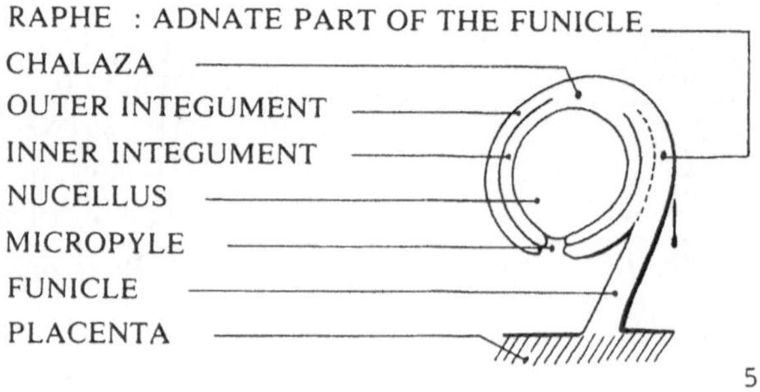

RAPHE : ADNATE PART OF THE FUNICLE

CHALAZA

OUTER INTEGUMENT

INNER INTEGUMENT

NUCELLUS

MICROPYLE

FUNICLE

PLACENTA

5

Plate 2

OVULES: SHAPE

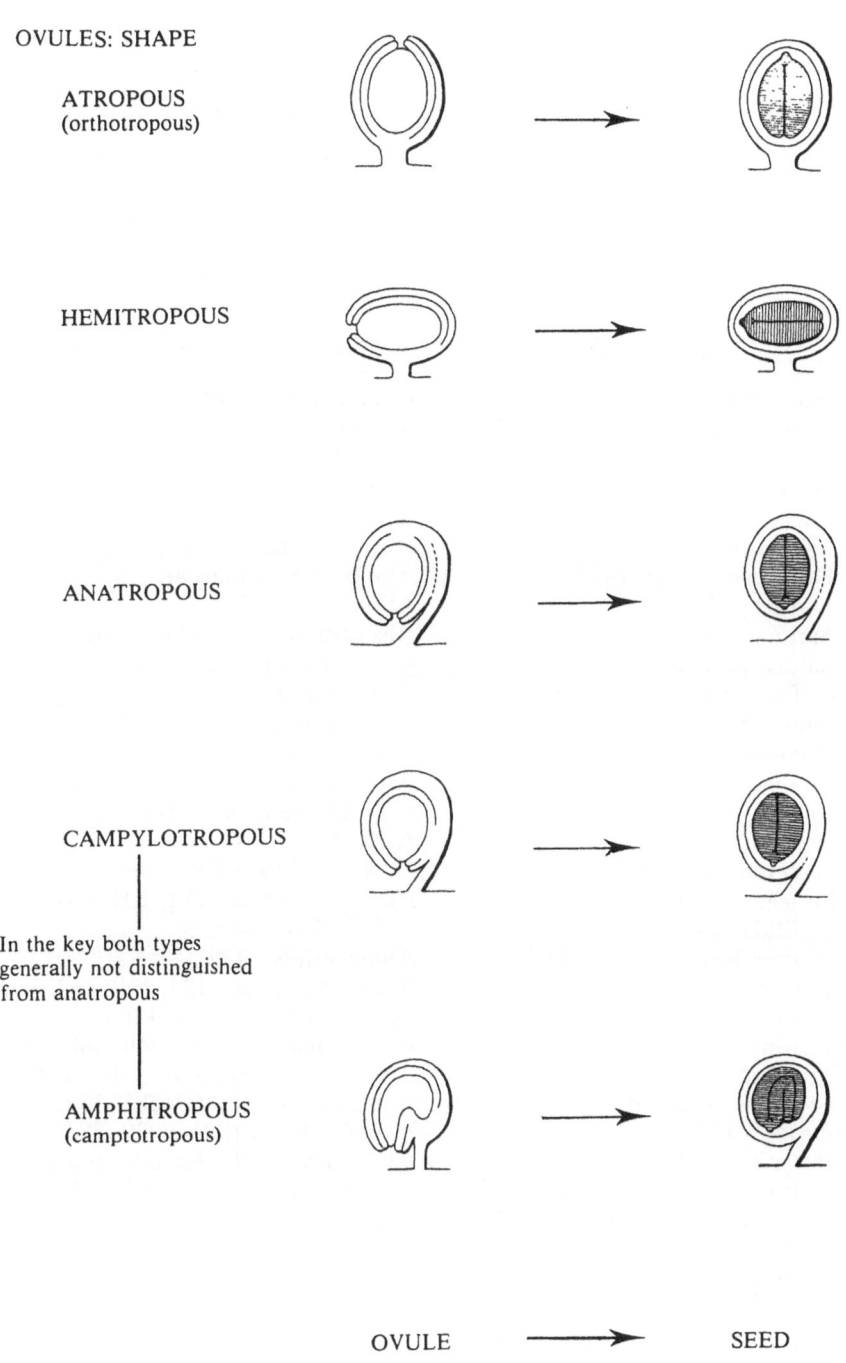

ATROPOUS
(orthotropous)

HEMITROPOUS

ANATROPOUS

CAMPYLOTROPOUS

In the key both types
generally not distinguished
from anatropous

AMPHITROPOUS
(camptotropous)

OVULE   ⟶   SEED

Plate 3

INDEX

Taxa below the rank of family mentioned in a lead may well key out elsewhere also, without being enumerated there, e.g. because too many taxa
to be noted are represented in that particular lead.
  The numbers cited refer to the number of the couplet, not page.

Basellaceae 1844
Batidaceae 206
Bauera 1069
Baueraceae
(= Saxifragaceae) 1069, 1147,
1277, 1302, 1567
Beesia 317
Begoniaceae 504, 538, 1555, 1563,
2049
Begoniella 2049
Bellendena 300
Belliolum 604
Bembicia 506
Berberidaceae 222, 557, 563, 564,
579, 593, 605, 819, 955
Berberis 557, 564
Berenice 2057
Berzelia 1400, 1494, 2036
Betulaceae 211, 212, 490, 522
Biebersteiniaceae
(= Geraniaceae) 380, 656
Bignoniaceae 1670, 1745, 1747,
1916, 1929, 1941, 1974, 1981,
1991
Bishofiaceae
(= Euphorbiaceae) 386
Bixaceae 829
Blepharocaryaceae
(= Anacardiaceae) 940
Bocconia 247
Boerlagellaceae[1]
Bomarea 133, 157
Bombacaceae 370, 699, 745, 852,
864, 871, 882, 1247, 1316, 1344,
1374, 1454, 1854, 1855, 1887,
1891, 2002, 2067, 2078
Bonnetiaceae (= Theaceae) 886,
911, 1653
Boraginaceae 1658, 1676, 1722,
1723, 1736, 1793, 1794, 1800,
1809, 1826, 1828, 1838, 1902,
1927, 1935, 1992
Bouea 560, 940
Brachynema 1857
Brasenia 921
Bretschneideraceae 1243
Brexia 1065

Brexiaceae
(= Saxifragaceae) 1065, 1207,
1253, 1605
Bromeliaceae 90, 148, 157
Bromelioideae 157
Brunelliaceae 452
Bruniaceae 1289, 1294, 1400,
1416, 1428, 1442, 1463, 1494,
1510, 1739, 1827, 2036, 2042
Brunoniaceae 1686, 1952
Bubbia 604, 1576
Buddleja 1792
Buddlejaceae
(= Loganiaceae) 1792
Burmanniaceae 16, 112, 121
Burmannieae 112
Burnatia 69
Burseraceae 684, 1039, 1049,
1057, 1230, 1239, 1628
Butomaceae 103, 149
Butomus 103
Buxaceae 352, 392
Byblidaceae 1607, 1756

Cabomba 788
Cabombaceae
(= Nymphaeaceae) 447, 454,
788, 921
Cactaceae 1545, 2090
Cadellia 804
Caesalpiniaceae, see Leguminosae
Calectasiaceae
(= Xanthorrhoeaceae) 99
Callianthemum 449
Callitrichaceae 189
Calostemma 124
Caltha 454

1 Boerlagellaceae are not included in
the key, as the family is based upon
incomplete material;
Boerlagella = Planchonella (Sapotaceae)
and **Dubardella = Pyrenaria**
(Theaceae).

216

Calycanthaceae   16, 1308, 1323,
   1364, 1365, 1367
Calyceraceae   2036
Camelliaceae, see Theaceae
Campanulaceae   538, 1460, 1602,
   1977, 1983, 2044, 2054, 2055,
   2057
Camptotheca   1472
Canacomyrica   221, 237
Canarium   1049
Canella   610
Canellaceae   610, 830, 1640
Cannabidaceae   271
Cannaceae   108
Canotia   1259
Canotiaceae   700
Cansjera   256, 1842
Capparaceae   322, 400, 566, 615,
   671, 692, 711, 719, 838, 841, 862,
   878, 893, 900, 933, 961, 1036,
   1064, 1092, 1096, 1116, 1120,
   1122, 1140, 1329, 1333, 1625,
   1997
Caprifoliaceae   2008, 2009, 2017,
   2046, 2065, 2072
Carallia   1408
Cardiopteridaceae   1693, 1824
Caricaceae   723, 762, 1875
Carlemanniaceae
   (= Caprifoliaceae)   2017
Carpinaceae (= Betulaceae)   522
Cartonemataceae
   (= Commelinaceae)   145
Carya   171, 202
Caryocar   855
Caryocaraceae   855, 1892
Caryophyllaceae   248, 264, 290,
   315, 473, 595, 726, 731, 954,
   1155, 1169, 1182, 1270, 1704
Casearia   295
Casuarinaceae   200
Cassipourea   1228, 1346
Cassytha   282, 2105
Castela   1111
Catophractes   1981
Cadrelopsis   748, 754

Celastraceae   310, 632, 663, 973,
   1015, 1022, 1028, 1029, 1044,
   1054, 1071, 1082, 1092, 1140,
   1162, 1169, 1175, 1182, 1200,
   1206, 1207, 1215, 1218, 1227,
   1236, 1238, 1258, 1259, 1286,
   1296, 1409, 1427, 1445, 1452,
   1587, 1602, 1700, 1729, 1771
Celosia   297
Celosieae   306, 314
Centrolepidaceae   33, 34
Centroplacus   1042
Cephalotaceae   434
Cephalotaxaceae   10
Ceratiosicyos   1590
Ceratopetalum   543
Ceratophyllaceae   274
Cercidiphyllaceae   178, 207, 456
Ceriops   1408
Ceropegia   1963
Chamaelaucieae   1418
Chenopodiaceae   204, 266, 473
Chimonanthus   1308
Chlaenaceae, see Sarcolaenaceae
Chloantheae   1939
Chloranthaceae   170, 219, 476
Chrysobalanaceae   294, 297, 1162,
   1170, 1199, 1324, 1339
Chunia   188
Cinnamodendron   830
Cinnamosma   1640
Circaea   1431
Circaesteraceae
   (= Ranunculaceae)   276, 432
Cistaceae   600, 610, 837, 844, 875,
   885, 907, 1118, 1123, 1148
Clavija   1849
Claytonia   16
Clematis   428
Clematoclethra   718
Cleomaceae (= Capparaceae)   566,
   671, 838, 961, 1120
Clethraceae   708, 769
Cneoraceae   1051
Cobaeaceae
   (= Polemoniaceae)   1765
Coccolobeae   728

Nelumbo 1
Nelumbonaceae
   (= Nymphaeaceae)   1362
Nemopanthus   358
Neoluederitzia   980
Neopringlea   311
Neostrearia   1513
Neotessmannia   1536
Nepenthaceae   400
Nesogordonia   868
Nestegis   216
Nettoa   840, 1122
Neumanniaceae
   (= Flacourtiaceae)   333
Neuradaceae   1490
Neuradoideae   1490
Nigella   413
Nigelleae   895
Nitraria   859, 1136
Nolanaceae   1778, 1803
Norantea   1869, 1885
Nyctaginaceae   254, 257, 1582,
   1616, 1637
Nyctocalos   1747
Nymphaeaceae   447, 454, 788, 896,
   921, 1109, 1352, 1362, 1538,
   1549, 1571, 2102
Nypaceae (= Palmae)   37
Nyssaceae   486, 1402, 1472, 1524

Oceanopapaver   841, 1122
Ochnaceae   625, 637, 643, 672,
   707, 828, 859, 884, 915, 1110
Ochradenus   323
Octoknemataceae
   (= Olacaceae)   498, 519, 1381
Octolepis   1606
Octolobus   429
Olacaceae   175, 498, 519, 573, 591,
   646, 813, 858, 932, 943, 953, 966,
   979, 990, 1002, 1055, 1129, 1175,
   1178, 1212, 1381, 1410, 1673,
   1846, 1857, 1873, 1883, 2070
Oldfieldia   386
Oleaceae   175, 216, 380, 385, 682,
   1661, 1730
Oligomeris   1945

Oliniaceae   1382, 1459
Onagraceae   540, 1203, 1390, 1417,
   1431, 1442, 1462, 1538, 1552
Oncothecaceae   1837
Ongokea   953
Ophiocaryon   689, 1047
Ophiopogon   128
Ophiopogonoideae   79
Opiliaceae   16, 242, 252, 256, 279,
   939, 1687, 1842
Orchidaceae   16, 77, 109, 152
Oresitrophe   330
Orobanchaceae   1914, 1958, 2114
Oroxylum   1747
Orygia   1571, 1575
Oxalidaceae   756, 764, 803, 895,
   1597, 1632
Oxygyne   112
Oxystylidaceae   671

Paeonia   1369
Paeoniaceae   1369
Pakaraimaea   887
Palmae   37
Panax   1499
Panda   981
Pandaceae s.s. (only Panda)   981;
   (rest see Euphorbiaceae)   1042
Pandanaceae   20, 53
Papaveraceae   16, 247, 262, 331,
   332, 565, 577, 608, 609, 735, 832,
   890, 1192, 1334, 1998
Papilionaceae, see Leguminosae
Parabaena   783
Paracryphiaceae   227, 410
Parinari   1199, 1339
Parnassiaceae
   (= Saxifragaceae)   1190, 1276,
   1483
Paronychioideae   473
Passiflora   327
Passifloraceae   327, 333, 620, 734,
   963, 965, 1188, 1274, 1325
Pauridia   114
Pedaliaceae   1908, 1928, 1936,
   1943, 2008
Peganum   881, 1142

225